MEASURING THE 21st CENTURY SCIENCE AND ENGINEERING WORKFORCE POPULATION

EVOLVING NEEDS

Panel to Evaluate the National Center for
Science and Engineering Statistics Approach to
Measuring the Science and Engineering Workforce

Committee on National Statistics

Division of Behavioral and Social Sciences and Education

A Consensus Study Report of

The National Academies of
SCIENCES · ENGINEERING · MEDICINE

THE NATIONAL ACADEMIES PRESS
Washington, DC
www.nap.edu

THE NATIONAL ACADEMIES PRESS 500 Fifth Street, NW Washington, DC 20001

This activity was supported by contracts between the National Academy of Sciences and the National Science Foundation (#SES-1024012). Any opinions, findings, conclusions, or recommendations expressed in this publication do not necessarily reflect the views of any organization or agency that provided support for the project.

International Standard Book Number-13: 978-0-309-46913-5
International Standard Book Number-10: 0-309-46913-9
Digital Object Identifier: https://doi.org/10.17226/24968

Additional copies of this publication are available for sale from the National Academies Press, 500 Fifth Street, NW, Keck 360, Washington, DC 20001; (800) 624-6242 or (202) 334-3313; http://www.nap.edu.

Printed in the United States of America

Suggested citation: National Academies of Sciences, Engineering, and Medicine. (2018). *Measuring the 21st Century Science and Engineering Workforce Population: Evolving Needs*. Washington, DC: The National Academies Press. doi: https://doi.org/10.17226/24968.

PANEL TO EVALUATE THE NATIONAL CENTER FOR SCIENCE AND ENGINEERING STATISTICS APPROACH TO MEASURING THE SCIENCE AND ENGINEERING WORKFORCE

RITA COLWELL (*Cochair*), Department of Cell Biology and Molecular Genetics, University of Maryland and Bloomberg School of Public Health, Johns Hopkins University
JAMES S. HOUSE (*Cochair*), Institute for Social Research, University of Michigan
JENNIFER SUE BOND, Council on Competitiveness
GEOFF DAVIS, Google, Inc.
DONALD DILLMAN, Department of Sociology, Washington State University
RICHARD FREEMAN, Department of Economics, Harvard University
JACK GAMBINO, Statistics Canada
MARESI NERAD, College of Education, University of Washington
RANDALL OLSEN, Department of Economics, Ohio State University
WILLIE PEARSON, JR., School of History, Technology, and Society, Georgia Institute of Technology
KEITH RUST, Westat
NORA CATE SCHAEFFER, Department of Sociology, University of Wisconsin–Madison
JAMES WAGNER, Institute for Social Research, University of Michigan
YU XIE, Department of Sociology, Princeton University

KRISZTINA MARTON, *Study Director*
ANTHONY MANN, *Program Coordinator*

Acknowledgments

Personally, and on behalf of the Panel to Evaluate the National Center for Science and Engineering Statistics Approach to Measuring the Science and Engineering Workforce, we wish to thank all those who contributed their time and expertise to inform the panel's work. The study was initiated by the National Science Foundation's (NSF) National Center for Science and Engineering Statistics (NCSES) under the guidance of director John Gawalt and deputy director Emilda Rivers. The panel would like to thank the NSF staff who assisted the panel with presentations and background materials, including Mark Fiegener, Beethika Khan, and Kelly Phou from NCSES and Matthew Wilson from the National Science Board Office. Throughout the course of the study, the panel especially benefited from the able and very generous assistance provided by John Finamore, program director of the NSF Human Resources Statistics Program.

The panel's understanding of the needs of data users was greatly expanded by the presentations at a public workshop of William Bonvillian, Massachusetts Institute of Technology; Lisa Frehill, Energetics Technology Center; Fernando Galindo-Rueda, OECD, France; Donna Ginther, University of Kansas; William Harris, Science Foundation Arizona; Kaye Husbands Fealing, Georgia Institute of Technology; Janet Metcalfe, Vitae, United Kingdom; Leigh Ann Pennington, Oak Ridge Institute for Science Education; Michael Roach, Cornell University; Neil Ruiz, George Washington University; Beate Scholz, independent consultant, Germany; John Skrentny, University of California, San Diego; Tobin Smith, Association of American Universities; Andrea Stith, BioFrontiers, University of Colorado Boulder; Prasanna Tambe, New York University; Yu Tao, Stevens Institute of Technology; Michael Teitelbaum, Harvard University; and Brian Yoder, American Society for Engineering Education.

The panel's work was supported by the capable staff of the National Academies of Sciences, Engineering, and Medicine. Brian Harris-Kojetin, director, and Connie Citro, former director, Committee on National Statistics, provided institutional leadership. Krisztina Marton superbly directed the study. Kirsten Sampson-Snyder, Division of Behavioral and Social Sciences and Education, oversaw the report review process. The report also greatly benefited from the meticulous editing provided by Rona Briere of Briere Associates, Inc. We also thank program associate Anthony Mann for very friendly and efficient logistical support for the panel's meetings.

Finally, and most important, we greatly appreciated the work of our fellow panel members, whose dedication, patience, and commitment made this report possible: Jennifer Bond, Council on Competitiveness; Geoff Davis, Google, Inc.; Don Dillman, Washington State University; Richard Freeman, Harvard University; Jack Gambino, Statistics Canada; Maresi Nerad, University of Washington; Randy Olsen, Ohio State University; Willie Pearson, Georgia Institute of Technology; Keith Rust, Westat; Nora Cate Schaeffer, University of Wisconsin–Madison; James Wagner, University of Michigan; and Yu Xie, Princeton University. Panel members were chosen for their expertise, subject matter knowledge, and varied perspectives, serving on the panel as individuals, not as representatives of their employers or organizations. They were a pleasure to work with, increasingly becoming a very intellectually and scientifically engaged and opinionated group, and a quite cohesive one as well.

This Consensus Study Report was reviewed in draft form by individuals chosen for their diverse perspectives and technical expertise. The purpose of this independent review is to provide candid and critical comments that will assist the National Academies of Sciences, Engineering, and Medicine in making each published report as sound as possible and to ensure that it meets the institutional standards for quality, objectivity, evidence, and responsiveness to the study charge. The review comments and draft manuscript remain confidential to protect the integrity of the deliberative process.

We thank the following individuals for their review of this report: Norman M. Bradburn, senior fellow at NORC at the University of Chicago; Patrick J. Cantwell, Decennial Statistical Studies Division, U.S. Census Bureau; Bonnie J. Dunbar, Aerospace Engineering, TEES Institute for Engineering Education and Innovation, Texas A&M; John L. Eltinge, Research and Methodology, U.S. Census Bureau; Donna K. Ginther, Department of Economics, University of Kansas; Richard C. Larson, Institute for Data, Systems, and Society, Massachusetts Institute of Technology; Shirley M. Malcom, Directorate for Education and Human Resources Programs, American Association for the Advancement of Science; Kristen Olson,

Department of Sociology, University of Nebraska–Lincoln; and Brady T. West, Survey Methodology Program, Institute for Social Research, University of Michigan.

Although the reviewers listed above provided many constructive comments and suggestions, they were not asked to endorse the conclusions or recommendations of this report, nor did they see the final draft before its release. The review of this report was overseen by report review coordinator V. Joseph Hotz, Department of Economics, Duke University, and report review monitor Kenneth Wachter, Departments of Demography and Statistics (emeritus), University of California, Berkeley. They were responsible for making certain that an independent examination of this report was carried out in accordance with the standards of the National Academies and that all review comments were carefully considered. The quality of the report has been enhanced by the entire review process. Responsibility for the final content rests entirely with the authoring committee and the National Academies.

Rita Colwell (*Cochair*), University of Maryland and Johns Hopkins University
James S. House (*Cochair*), University of Michigan
Panel to Evaluate the National Center for Science and Engineering Statistics Approach to Measuring the Science and Engineering Workforce

Acronyms and Abbreviations

ACS	American Community Survey
CARRA	Census Bureau's Center for Administrative Records Research and Applications
ECDS	Early Career Doctorates Survey
ED	U.S. Department of Education
EIN	employer identification number
FFRDC	Federally Funded Research and Development Center
FSRDC	Federal Statistical Research Data Center
GSS	Survey of Graduate Students and Postdoctorates in Science and Engineering
IRS	Internal Revenue Service
LEHD	Longitudinal Employer-Household Dynamics Program
NASA	National Aeronautics and Space Administration
NCES	National Center for Education Statistics
NCSES	National Center for Science and Engineering Statistics
NIH	National Institutes of Health
NIH IRP	National Institutes of Health Intramural Research Program
NSCG	National Survey of College Graduates

NSF	National Science Foundation
NSRCG	National Survey of Recent College Graduates
SDR	Survey of Doctorate Recipients
SED	Survey of Earned Doctorates
SEI	*Science and Engineering Indicators* report
S&E	science and engineering
SSA	Social Security Administration
UI earnings	unemployment insurance earnings
USDA	U.S. Department of Agriculture
WMPD	*Women, Minorities, and Persons with Disabilities in Science and Engineering* report

Contents

Figures, Tables, and Boxes

FIGURES

TABLES

BOXES

Summary

Reflecting the value of science and engineering in World War II and the Cold War, the United States created during the 1950s, 1960s, and 1970s a capacity within the National Science Foundation (NSF) to collect, analyze, and disseminate information on the nation's science and engineering workforce. The America COMPETES Act of 2010 reaffirmed these efforts and consolidated them within the National Center for Science and Engineering Statistics (NCSES) at NSF. NCSES is one of the principal statistical agencies in the United States and a leading provider of data on the nation's science and engineering workforce. These data derive heavily from long-running surveys, including the Survey of Graduate Students and Postdoctorates in Science and Engineering, the Survey of Earned Doctorates, the Survey of Doctorate Recipients (SDR), the National Survey of College Graduates (NSCG), and a new Early Career Doctorates Survey (ECDS). The data inform congressionally mandated reports and are used extensively by policy makers, researchers, industry, the media, and others.

The NCSES surveys have represented a gold standard for data in this area, not only in the United States but also internationally, as the result of an active program of evaluation and improvement. In this vein, and responding to the rapidly changing nature of society and the science and engineering workforce, NCSES asked the National Academies of Sciences, Engineering, and Medicine to convene a panel to "review, assess, and provide guidance on NCSES's approach to measuring the science and engineering workforce . . . [and] to develop a framework . . . that provides flexibility to examine emerging issues . . . while . . . allowing for stability in the estimation of key trends."

The panel identified several priority areas of focus for NCSES to ensure that the data it provides remain robust and relevant in the coming decades. These priority areas are highlighted below. The report contains additional recommendations focused on further strengthening the NCSES science and engineering workforce surveys. The panel acknowledges that planning and implementing the building blocks of a new framework will require substantial investments of time and resources.

1. The increasing fluidity and complexity of the career pathways of scientists and engineers necessitate placing increased focus and generating longitudinal data on the nature, determinants, and consequences of these pathways.

   RECOMMENDATION 3-1: The National Center for Science and Engineering Statistics should place greater emphasis on following sample members longitudinally and obtaining survey data on the nature, determinants, and consequences of significant transitions in science and engineering career pathways.

   RECOMMENDATION 4-1: The National Center for Science and Engineering Statistics should maintain and accelerate its efforts to develop sample designs that facilitate the increased generation and analysis of longitudinal data in the National Survey of College Graduates and the Survey of Doctorate Recipients, including the generation of appropriate weights and documentation for longitudinal analysis.

   RECOMMENDATION 6-6: The National Center for Science and Engineering Statistics should facilitate the presentation of results of longitudinal analyses in the *Science and Engineering Indicators* and *Women, Minorities, and Persons with Disabilities in Science and Engineering* reports.

2. Science and engineering career pathways increasingly involve international mobility.

   RECOMMENDATION 3-2: The National Center for Science and Engineering Statistics should work toward survey design and questionnaire content that capture both the immigration of individuals from abroad into the U.S. science and engineering workforce and the emigration of individuals from the U.S. science and engineering workforce to other countries, including the nature, determinants, and employment outcomes of such movements. Specific questionnaire modules targeted to these groups could explore such issues.

RECOMMENDATION 4-2: The National Center for Science and Engineering Statistics should evaluate whether modeling techniques could be used to improve the sensitivity of the classification used to stratify the National Survey of College Graduates (NSCG) sample drawn from the American Community Survey, thereby making it more efficient to increase in the NSCG the sample of doctorate holders who obtained their degree from abroad.

3. Fluid and complex career pathways create and are created by increasingly fluid and complex patterns in the acquisition of skills and training and the skills and training required in various jobs along the career pathway.

   RECOMMENDATION 3-3: To address the desire of researchers and policy makers for in-depth information on such topics as training and skills obtained and needed, the National Center for Science and Engineering Statistics should explore and evaluate adding to its surveys topic modules that would vary from round to round or be asked only of subsets of the samples.

4. The diversity of the workforce has been increasing rapidly, requiring increased attention to factors that facilitate or impede entry into and retention of groups historically underrepresented in the science and engineering workforce.

   RECOMMENDATION 3-5: The National Center for Science and Engineering Statistics should assess the performance of the new questions on working conditions and work–life balance in the Early Career Doctorates Survey and should consider adding these questions to the National Survey of College Graduates and the Survey of Doctorate Recipients as well.

   RECOMMENDATION 3-6: The National Center for Science and Engineering Statistics should develop for the surveys core questions and a more in-depth module on harassment and discrimination.

5. To facilitate achievement of these substantive goals will require that NCSES continue its strong program of methodological research, with increased focus in two areas.

   a. Accelerate research on the use of administrative and other non-survey data to complement or replace survey data, reduce respondent burden, and reduce data collection costs. Research on the feasibility of linkages to unemployment insurance earnings data may be most worthwhile to pursue.

RECOMMENDATION 3-7: The National Center for Science and Engineering Statistics should explore mechanisms for accelerating research on the use of alternative data sources to expand and supplement the information obtained from the science and engineering workforce surveys.

RECOMMENDATION 5-6: The National Center for Science and Engineering Statistics should evaluate the feasibility of using administrative data, such as unemployment insurance earnings data, to inform adaptive design, address nonresponse due to lack of contact information, conduct nonresponse bias analysis, and develop nonresponse weights.

b. Continue ongoing research aimed at enhancing the cost effectiveness of the surveys.

RECOMMENDATION 5-1: The National Center for Science and Engineering Statistics (NCSES) should continue to research the optimal mix and sequence of contact, response modes, and incentives. Lessons learned from the National Survey of College Graduates should be evaluated in the context of the other surveys, particularly the Survey of Doctorate Recipients. NCSES should ensure that costs are tracked in sufficient detail, for both new and continuing participants, to enable sophisticated comparisons of the different options, along with the implications for data quality.

RECOMMENDATION 5-8: The National Center for Science and Engineering Statistics should evaluate the data processing methods used for the surveys to identify potential opportunities to increase efficiency and the timeliness of data releases.

6. Continuing effort on the part of NCSES is needed to facilitate the dissemination of (1) reports and other information/data products and (2) microdata, especially longitudinal microdata files.

RECOMMENDATION 6-4: The National Center for Science and Engineering Statistics should continue making its work toward enabling longitudinal research based on the survey data a priority and should seek input from other agencies that provide access to longitudinal data.

RECOMMENDATION 6-5: The National Center for Science and Engineering Statistics should continue to emphasize the timeliness of its data releases and the provision of clear documentation of the data and should consider hosting an online forum or providing some other mechanism that can facilitate discussion among users of the survey data.

RECOMMENDATION 6-7: The National Center for Science and Engineering Statistics should evaluate options for expanding communications with former, existing, and new stakeholders to raise awareness of the usefulness of the survey data. State and local governments warrant special consideration, as do groups historically underrepresented in the science and engineering workforce.

1

Introduction and Framework for the Future

The National Science Foundation's (NSF) National Center for Science and Engineering Statistics (NCSES), one of the nation's principal statistical agencies, is charged[1] to "collect, acquire, analyze, report, and disseminate statistical data related to the science and engineering enterprise in the United States and other nations that is relevant and useful to practitioners, researchers, policy makers, and to the public, including statistical data on":

- research and development (R&D) trends;
- the science and engineering workforce (the focus of the present study);
- U.S. competitiveness in science, engineering, technology, and R&D; and
- the condition and progress of U.S. science, technology, engineering, and mathematics (STEM) education.

Like its predecessors, NCSES organizes the data it collects and acquires into reports and provides the data (with appropriate ethical and confidentiality protections) to researchers and policy makers for their analysis. Thus, NCSES data have become the major evidence base for American science and

[1] In 2010, Section 505 of the reauthorization of the America COMPETES Act established a National Center for Science and Engineering Statistics as 1 of the 13 federal statistical agencies, located within NSF and absorbing the prior functions of the Division of Science Resources Statistics (DSRS).

technology policy, and by all accounts the agency is well respected globally for these data.

NCSES's past and current high standing has been maintained in part through a continuing process of evaluation and improvement, in which reports produced by panels of the National Academies of Sciences, Engineering, and Medicine at the request of NCSES have played a critical role (see Chapter 2). The most recent of these panels was asked by NCSES "to assess and provide recommendations regarding the need for revised, refocused, and newly developed indicators designed to better reflect fundamental and rapid changes that are shaping global science, technology, and innovation systems" (National Research Council, 2014, p. 6). One chapter of that panel's report focuses on "measuring human capital," which is the exclusive focus of the present report. Given the centrality of the measurement of the science and engineering workforce to NCSES's mission, activities, staff, and budget, NCSES and NSF requested that the National Academies' Committee on National Statistics "convene an ad hoc panel to review, assess, and provide guidance on NCSES's [current] approach to measuring the science and engineering (S&E) workforce population in the United States," . . . [especially] "to develop a framework for measuring the S&E workforce [into the future] that provides flexibility to examine emerging issues related to this unique population while at the same time allowing for stability in the estimation of key trends." The full charge to the panel is given in Box 1-1.

A suite of surveys constitutes the core of NCSES's efforts to characterize the constantly changing nature of the science and engineering workforce. These surveys have evolved and improved over half a century. The panel's overall approach to addressing its statement of task was to consider the data collection, analysis, and reporting systems to which NCSES should be aspiring and moving in the future while also providing guidance on how to make the current system more efficient and how to transition toward a framework that can be implemented in the longer term.

HISTORICAL CONTEXT

During and after World War II, the role and value of the nation's scientists and engineers and their research were recognized as important to securing victory in the war. The need to support and monitor the science and engineering enterprise came into increased focus with the advent of the Korean War. The deficit of trained scientists and engineers resulting from World War II was a primary concern addressed by two major reports of the time: *Science: The Endless Frontier* by Vannevar Bush (1945) (reprinted by the National Science Foundation in 1990) and the Steelman report (Steelman, 1949).

BOX 1-1
Statement of Task

The National Academies of Sciences, Engineering, and Medicine's Committee on National Statistics, at the request of the National Science Foundation's, National Center for Science and Engineering Statistics (NCSES), will convene an ad hoc panel to review, assess, and provide guidance on NCSES's approach to measuring the science and engineering (S&E) workforce population in the United States. Given the evolving data needs of NCSES stakeholders and the budget climate uncertainty under which NCSES operates, a priority for the panel is to develop a framework for measuring the S&E workforce that provides flexibility to examine emerging issues related to this unique population while at the same time allowing for stability in the estimation of key trends.

The framework developed by the panel should provide direction on measuring the S&E workforce population in terms of content, data sources, survey design, survey methodology, data collection, data processing, data integration, and data dissemination. The panel should consider the challenges involved in: measuring the S&E workforce over multiple decades, using a variety of data sources given that the availability of these sources could change over time, taking advantage of lessons learned from how other developed countries collect information on scientists and engineers, employing the most efficient and effective methodology at any given time, and meeting the evolving data needs of policy makers, researchers, and the public.

At the end of its review, the panel will issue a report with findings, recommendations, and priorities for improving the relevance, accuracy, timeliness, and cost effectiveness of S&E workforce data for the next decade and beyond.

In recognition of the foundational importance of science and engineering to the fundamental goals of the U.S. government, President Truman signed into law the National Science Foundation Act of 1950. The law gave NSF the mandate "to promote the progress of science; to advance the national health, prosperity, and welfare; to secure the national defense; and for other purposes" (Public Law 81-507). As one adjunct of the central goal "to develop and encourage the pursuit of a national policy for the promotion of basic research and education in the sciences" (including engineering, medicine, and other more applied sciences), NSF was to "maintain a register of scientific and technical personnel and in other ways to provide a central clearinghouse for information covering all scientific and technical personnel in the United States."

In September 1950, William T. Golden was charged by the Bureau of the Budget to prepare a report for President Truman on how the nation might best mobilize its scientific resources for a possible protracted military emergency. He interviewed the country's top scientists and military officers before recommending that the president appoint a full-time science advisor. President Truman subsequently adopted the essence of this recommendation by establishing the Scientific Advisory Committee to the White House Office of Defense Mobilization, which later evolved into a full-scale presidential scientific advisory system. Golden also was instrumental in supporting the establishment of NSF. He provided a memorandum outlining a possible program for the agency that included four recommendations, two of which related to science and technology personnel:

- Golden's second recommendation called for a "survey detailing the existing support of graduate and undergraduate education in the sciences by the many public and private agencies so engaged," and noted that "a clear understanding of the existing situation is a condition precedent to an effective fellowship and scholarship program by the National Science Foundation."
- Golden's third recommendation called on NSF to "study the scientific manpower resources of the United States: (a) as specifically called for in the Act, by taking over, completing, and keeping current the detailed National Scientific Register; and (b) by preparing quantitative analytical studies of available and prospective scientific and technical manpower. The categories would include scientific and technical specialties as well as degrees of proficiency, years of experience, age brackets and the like. These latter studies would be based in part on judgment and statistical analyses of available data rather than just on head counts. Dependable data of this sort is badly needed in mobilization planning and would influence Selective Service policies and decisions as to educational emphasis and support by Government and private agencies." (Blanpied, 1995)

NSF's third annual report, covering the period July 1, 1952, to June 30, 1953, included the first survey results on human resources for science and engineering. The report was well received by President Eisenhower, who issued an Executive Order (10521) requiring NSF to continue to produce such comprehensive reports. Thus, the Division of Science Resources Statistics (DSRS) was established within NSF early in its history.

The Cold War and the launching of Sputnik I by the Soviet Union highlighted the continuing need for science and engineering resources and information about those resources. In 1972, the National Science Board concluded that it could provide a significant service to the nation by issuing a series of reports on science indicators based on quantitative data, including data on science and technology personnel. In 1982 and again in 1987, Congress mandated biennial *Science and Engineering Indicators*

(SEI) reports (see National Science Board [2000] for a detailed history of the importance of quantitative data, including data and analyses on science and engineering human resources, during this period).

The *SEI* reports are produced by the National Science Board. These are highly regarded flagship reports on all quantitative data regarding the U.S. science and engineering enterprise in a global context. They are authoritative, policy neutral, and extensively reviewed, and over the years have reported on a wide variety of data on science and engineering personnel. The reports take a broad look at the science and engineering landscape, ranging from elementary and secondary mathematics and science education to labor market conditions and beyond, relying on data from a variety of sources. The 2016 *SEI* report contains two chapters focused on higher education in science and engineering and the science and engineering labor force. The chapters on academia and industry also contain salient information, for instance, on the doctoral and postdoctoral science and engineering workforce. These chapters rely primarily on the various NSF human resource surveys detailed below but also include data from other external sources. Appendix A summarizes the *SEI* figures and tables that rely on data from the NCSES science and engineering surveys that are the focus of this report. Because the *SEI* data and analyses are extensive, the reports have not been published in hard copy since 2016, but are fully available electronically (see National Science Board, 2016).

In 1980, Congress enacted the National Science Foundation Authorization and Science and Technology Equal Opportunities Act (Public Law 96-516). The act established within NSF a Committee on Equal Opportunities in Science and Technology to advise the agency concerning the act's implementation. The act also required the NSF director to prepare and transmit to Congress a biennial report concerning the participation and status of women and minorities in science and technology, including an accounting and comparison by sex, race, and ethnic group and by discipline of the participation of women and men in scientific and technical positions. This legislation was enacted in recognition of the importance not only of providing equal opportunities but also of expanding the diversity of people entering STEM fields. It reflects awareness that population shifts in the United States necessitate training and employing a greater percentage of women and underrepresented minorities in STEM fields if the nation is to recruit and employ an adequate science and engineering workforce and maintain its global competitive advantage.

According to the 2017 *Women, Minorities, and Persons with Disabilities in Science and Engineering* report, the representation of certain groups of people in science and engineering education and employment still differs from their representation in the U.S. population (National Center for Science and Engineering Statistics, 2017g). Women, persons with disabilities, and three racial/ethnic groups—blacks, Hispanics, and American Indians/Alaska Natives—are underrepresented in science and engineering. Women have reached parity with men among recipients of science and engineering degrees overall, but their representation differs by field, and they constitute disproportionately smaller percentages of employed scientists and engineers relative to their percentage of the U.S. population. Blacks, Hispanics, and American Indians/Alaska Natives have gradually increased their share of science and engineering degrees, although the extent of this increase varies across fields, while they have remained underrepresented overall in educational attainment and the science and engineering workforce (National Center for Science and Engineering Statistics, 2017g). It should be noted that sample size limitations in the Survey of Doctorate Recipients (SDR) have historically made it particularly difficult to conduct in-depth analysis on subgroups.

The need for and role of science and engineering human resources continue to represent an issue of national economic competitiveness and security. A report of the National Research Council on research universities and the future of the United States highlights the importance of these universities to the nation's prosperity and national security (National Research Council, 2012b). Among that report's recommendations is the need to focus on building the talent of all Americans, including women and underrepresented minorities, in STEM fields; reforming graduate education; and ensuring that the United States will continue to benefit from the participation of international students and scholars. Likewise, Pressman and colleagues (2017) document that from 1996 to 2015, academic patents and the subsequent licensing to industry increased the gross output of U.S. industry by up to $1.33 trillion and U.S. gross domestic product (GDP) by up to $591 billion, and supported up to 4.3 million person-years of employment (see also Roessner et al., 2017).

The remainder of this chapter provides a brief overview of the current NCSES surveys, which are described more fully in Chapter 2. It then describes the key elements of a framework for further improving NCSES's measurement of the science and engineering workforce, which are addressed more fully in subsequent chapters:

- evolving data needs (Chapter 3);
- sample design and estimation procedures (Chapter 4);
- data collection and processing (Chapter 5); and
- data analysis, dissemination, and outreach (Chapter 6).

NCSES'S CURRENT APPROACH TO MEASURING THE SCIENCE AND ENGINEERING WORKFORCE

At the core of NCSES's current approach to measuring the U.S. science and engineering workforce are two census surveys and two probability surveys, supplemented for selected purposes by data from other sources, including NCSES's own Business Research and Development and Innovations Survey (BRDIS), the Bureau of Labor Statistics' Occupation Employment Statistics (OES) Program, and the Office of Personnel Management's (OPM) database of the U.S. science and engineering workforce. The principal components of the NCSES data system, as shown in Table 1-1, are the (complete census) Survey of Graduate Students and Postdoctorates in Science and Engineering (GSS), the (complete census) Survey of Earned Doctorates (SED), the National Survey of College Graduates (NSCG), and the SDR. In addition, every 2–3 years NCSES also conducts the Survey of Postdocs at Federally Funded Research and Development Centers. A new Early Career Doctorates Survey (ECDS) also is currently being launched. All of the existing surveys are widely used as a source for NCSES reports, policy analysis, and basic research (National Center for Science and Engineering Statistics, 2017b, 2017c, 2017d, 2017e; National Academy of Sciences, National Academy of Engineering, and Institute of Medicine, 2010; National Research Council, 2012a).

The GSS, conducted since 1972, is an annual census of all U.S. academic institutions granting research-based master's or doctoral degrees in science, engineering, and health fields. From each institution, the GSS collects the total number of graduate students, postdoctoral appointees, and doctoral-level nonfaculty researchers classified by gender and other characteristics, such as source of financial support. Aggregate data are reported in the fall of each year by institutional coordinators from all U.S. institutions granting research-based master's degrees or doctorates in science, engineering, and selected health fields.

The SED, an annual census of all persons receiving a research doctorate in any discipline from an accredited U.S. academic institution in a given academic year, is the oldest and best-established of the NCSES core surveys. It has been conducted since 1958, but data are available on individuals who received doctorates from U.S. institutions beginning in 1920.

The NSCG, which began in 1973, is the primary means by which NCSES collects information on the science and engineering workforce. The survey samples individuals who have at least a bachelor's degree in any discipline and has undergone a number of design revisions over the years, most recently when the American Community Survey (ACS) became the sampling frame for college graduates. The NSCG uses a rotating panel design that includes a new sample of graduates every biennial cycle. Sample members are

TABLE 1-1 Science and Engineering Education and Workforce Surveys of the National Center for Science and Engineering Statistics

Survey	Target Population	Focus of the Survey	Data Collection Frequency
Survey of Graduate Students and Postdoctorates in Science and Engineering (GSS)	Graduate students, postdoctoral appointees, and doctoral-level nonfaculty researchers at a U.S. academic institution granting research-based master's or doctoral degrees in science, engineering, and health fields[a] (information collected from institutional coordinators)	Graduate enrollment and postdoctoral appointees	Annual
Survey of Earned Doctorates (SED)	Recipients of research doctoral degrees in any field graduating from U.S. academic institutions	Doctoral degrees	Annual
National Survey of College Graduates (NSCG)	Individuals living in the United States who have at least a bachelor's degree in any field	Workforce	Biennial
Survey of Postdocs at Federally Funded Research and Development Centers (FFRDCs)	Postdocs employed at FFRDCs	Workforce	Periodic (every 2–3 years)
Survey of Doctorate Recipients (SDR)	Individuals with a science, engineering, or health research doctoral degree from a U.S. academic institution (sample based on the SED)	Workforce	Biennial
Early Career Doctorates Survey (ECDS)	Postdoctoral fellows and other doctoral degree recipients working at a U.S. academic institution, at an FFRDC, or in the National Institutes of Health Intramural Research Program who received a doctorate-equivalent degree in the past 10 years in any field and any country	Workforce	To be determined

[a] For a discussion of science and engineering degree fields, as well as scientists and engineers, see NCSES's Scientists and Engineers Statistical Data System Frequently Asked Questions at https://www.nsf.gov/statistics/sestat/#metadata&sestat-faq [October 23, 2017].

recontacted for three additional cycles (through 6 years postgraduation). This is NCSES's only source of detailed information on the education and occupations of the science and engineering workforce below the doctoral level (i.e., bachelor's and master's degree holders).

The SDR also began in 1973, when NCSES began to follow up a sample of respondents from the SED on a biennial longitudinal basis. A sample of new recipients of a U.S. research doctoral degree in a science, engineering, or health field is added at each cycle, and other new samples are added occasionally as a result of special requests or funding (as in 2015). Typically, sample members are followed until the age of 76. The number and balance of longitudinal cases versus new (cross-sectional) entries are determined at the time of each cycle (see Chapters 2 and 4 for additional detail on the sample designs). Since 2005, NCSES also has periodically conducted a Survey of Postdocs at Federally Funded Research and Development Centers (see Chapter 2).

Although the nature, strengths, and weaknesses of each of the surveys listed in Table 1-1 are discussed in more detail in Chapter 2, key aspects of these surveys are noted here. All are fixed-interval, largely cross-sectional surveys (or censuses), with some (again fixed-interval) longitudinal components. Aside from the derivation of the SDR sample from the SED, there are no data linkages over the separate surveys or with other major data sources. Finally, the content is focused on characterizing relatively detailed (or "fine") fields of degree and occupation, along with other demographics (e.g., gender, race/ethnicity, age) and on occupationally descriptive data (e.g., type of employment and employer, income/salary/wages).

The surveys have innovated and improved continually over the years in response to new substantive concerns and improvements in survey and other data collection methods and technologies. Yet they also still maintain some of their original design characteristics because of the need for caution in preserving trend data. They also reflect some of the important characterizations of and assumptions about the science and engineering workforce in the mid- to late 20th century: composed largely of native-born males who were sole or primary breadwinners, with doctoral-level scientists and engineers being a major driving force for scientific innovation and productivity, and with education and training occurring mainly in early adulthood and leading to fairly stable careers or jobs that reflected and utilized that education and training (National Academies of Sciences, Engineering, and Medicine, 2016a; National Science Board, 2015, 2017; Weisberg and Galinsky, 2014). The survey content also reflects much of the social, political, and economic context of the Cold War era in the United States and globally, as well as the organization of economic and technological innovation and development in universities, government, and large private organizations.

Increasingly since the 1980s, however, dramatic changes have occurred, and continue to do so, in all of the conditions and assumptions underlying the nature and functioning of the science and engineering workforce; the social, political, economic, and geographic context in which that workforce is situated and functions; and the sources and methods for collecting, organizing, analyzing, and reporting/disseminating data on the workforce. Thus, while respecting the need to maintain comparability in key data over time, NCSES and this panel are interested in a new framework for the science and engineering workforce surveys that reflects the realities of the nature and functioning of this workforce in the 21st century, as well as changes and opportunities in both (1) the nature and sources of data needed and available for describing and understanding the dynamically changing workforce, and (2) the methods for collecting, analyzing, reporting, and disseminating such data. The possibilities for and broad outline of a potential new framework are increasingly clear. Maintaining and enhancing the strengths of the current system while evolving it toward a new framework will be challenging, and that challenge is the focus of the succeeding chapters of this report. The remainder of this chapter describes the initial outline of this aspirational framework and ways of moving toward it.

AN ASPIRATIONAL FRAMEWORK FOR THE FUTURE

Evolving Data Needs (Chapter 3)

Several changes in the past decades in the nature of work and employment over the life course, especially for the science and engineering workforce, have yet to fully be reflected in the current suite of NCSES surveys. Perhaps most important is the increasing variability and fluidity of occupational and educational careers as a function of increasingly dynamic changes in the social and economic organization of society. Over just three decades, for example, information technology has become one of the dominant economic sectors and social forces in the United States and other societies (Auriol et al., 2013; National Research Council, 2012a). Individuals in the mid-20th century generally moved fairly linearly and steadily through educational levels and into jobs for which their education had prepared them, and then often had quite stable long-term careers within a given occupation and even work organization or employer. By contrast, current generations have much more irregular patterns of educational attainment, moving in, out of, and across higher education institutions and levels, as well as similarly variable patterns of employment that entail increasing rates of change across occupations and employers and employment sectors (e.g., academic, government, and private employers and, increasingly, self-employment/entrepreneurship) and even academic fields (National Science Board, 2015, 2017). Approaches to caring for family members (such as children and aging parents) while pursuing careers in science and engineering fields also have grown more diverse (Nerad, 2009).

To understand the sources, nature, and future directions of the science and engineering workforce at any given point in time, then, requires much better understanding of the nature, causes, and consequences of career pathways through which individuals move over their life course. To this end, there is a growing need for longitudinal data from the science and engineering workforce surveys. Although the NSCG and SDR follow sample members longitudinally, NCSES's focus in the past has been on producing cross-sectional estimates, and the data released have not been configured in ways that would facilitate longitudinal analysis (see Chapter 4). The panel believes that going forward, greater emphasis is needed on enabling and supporting use of the data for longitudinal analyses, which would expand the usefulness of these surveys in important ways. Longitudinal data that follow individuals through the pathways of their careers in higher education and the workforce can help researchers understand their reasons for changing those pathways and the precursors and antecedents of such changes. Educational and occupational pathways no longer occur primarily within nation states, but internationally and globally. Thus, NCSES data need especially to capture the patterns of immigration into American science and engineering educational and work organizations by foreign nationals, as well as the subsequent emigration to home countries or other nations of some of these immigrants and the emigration of native scientist and engineers. Similarly, it is important to understand whether and why members of the American science and engineering workforce work for American, foreign, or multinational organizations, and the consequences for both the workforce and scientific and technological development.

A major factor shaping the career paths of scientists and engineers, as well as their productivity and personal satisfaction, is the skills and training acquired in the course of education and work experience and their match with and utilization in current jobs. The increased fluidity of the occupational structure and of individual careers creates an increased need to assess and reassess both the skills and training individuals do and do not have and the skills and training required and used in their work settings (National Academy of Sciences, National Academy of Engineering, and Institute of Medicine, 2010).

The science and engineering workforce also is becoming increasingly diverse, not only in terms of nativity but also, within both native and foreign-born populations, in terms of gender, race/ethnicity, and other characteristics, although this increased diversity has occurred only gradually and modestly over the past few decades and has varied greatly by field and group (National Academy of Sciences, National Academy of Engineering, and Institute of Medicine, 2010; National Science Board, 2016). In particular, a wide gap in educational attainment and workforce participation still exits for women and underrepresented minorities in science and engineering fields such as engineering and computer science. The future recruitment, growth, and development of the nation's scientists and engineers will depend on greater understanding not only of the diverse composition of the science and engineering workforce but also of the factors that facilitate or impede the entry, retention, and advancement of underrepresented groups in the workforce. These factors range from marital and family status, to work–life balance, to conditions of work, to patterns of discrimination and harassment. NCSES needs to incorporate increased measurement of these and other emerging issues in all of its surveys.

In the future, it will potentially become more feasible to obtain a growing array of important data that can be used to understand the nature, causes, and consequences of various facets of the science and engineering workforce from administrative records or other forms of nonsurvey data. Such data include characteristics of organizations and geographic contexts in which scientists and engineers work, data on the products of their work (e.g., publications and patents), employment and earnings histories, and so on. NCSES has already begun to research the use of such administrative records and other nonsurvey data, and these sources will likely have an increasing role as part of the human resources data system in the future.

The U.S. workforce is becoming increasingly more diverse as well in terms of educational background. The SED and SDR have been the jewels in NCSES's crown for many decades, but capturing the science and engineering workforce across all postsecondary levels is becoming more important. Whole science and technology enterprises, most notably the information technology sector, are developed and led by people with only a bachelor's or master's degree (or even no postsecondary degree, a segment of the workforce not captured by the current surveys, in part because of sample design constraints). Greater emphasis on such individuals, and even those with specialty training and degrees from community colleges, is therefore imperative (see National Academies of Sciences, Engineering, and Medicine, 2016a).

Sample Design and Estimation Procedures (Chapter 4) and Data Collection and Processing (Chapter 5)

The content needs described above, together with developments in survey and data methods, suggest a potential future framework for the design of the science and engineering workforce survey system with several important characteristics.

First, as suggested above, a new framework would make greater use of nonsurvey data resources, including administrative data, although surveys would remain an essential aspect of the data system. These nonsurvey data sources could be useful in developing sampling frames, validating survey data, improving nonresponse weighting and imputation, substituting for some survey data to reduce respondent burden, and adding contextual or supplementary data to individual-level survey data.

Second, an integrated and longitudinal system of surveys and ancillary data would make the data collected more useful. Currently the SED and SDR have no relation to or overlap with the ECDS or NSCG. The ECDS could potentially be incorporated into the SDR, providing comparable national data on the same or similar topics now included only in the ECDS and allowing follow-up of ECDS samples over their life course. Following NSCG sample members longitudinally for longer periods throughout their life course would allow understanding of career pathways from the baccalaureate through doctoral levels and moves across those educational levels, as well as the career pathways of members of the science and engineering workforce with bachelor's and master's degrees. Although there are major practical limitations to developing a sampling frame based on institutional or administrative records for the NSCG (similar to the SED in the case of the SDR), the tradeoffs of different methods are constantly evolving and need to be researched continually. Such a shift may ultimately be necessary should the current ACS sampling frame cease to exist or become inadequate or inaccessible for NCSES's needs.

Finally, technological advances in the digital age, combined with the growth of alternative data sources, open up new and innovative methods for the collection of survey data. In ongoing research, it will be important to maintain participation rates and data quality as the circumstances of sample members' lives and their interactions with technology change. The NCSES surveys are increasingly moving to multimode methods (e.g., combinations of mail, Web, telephone, and face-to-face data collection), but much more use of such multimode methods will be possible in the future. In particular, in the future, administrative data sources may be able to substitute for and improve upon self-reports for the collection of certain kinds of data. One of the most promising data sources may be unemployment insurance earnings records, which also can be linked to the Quarterly Census of Employment and Wages. These data could in turn cue brief and specific follow-up survey modules with members of the NCSES longitudinal panels regarding, for example, the nature of and reason for unemployment, a job change, or a major change in income. These modules could reduce the frequency and intensity of fixed-interval longitudinal follow-up surveys and create a record of career pathways in real time. This change could also open up new possibilities for the use of modules tailored to and stimulated by survey responses or other data (e.g., job loss or change, marital and family change, reports of discrimination or harassment).

Data Analysis, Dissemination, and Outreach (Chapter 6)

A key problem for NCSES, as for any federal statistical agency, is how to get information and data to various users in a timely, well-documented, and targeted way. Such processes are always resource-constrained and must be prioritized, but advances in technology offer new options and possibilities. Were NCSES to achieve a more integrated, longitudinal, and timely mode of data collection (i.e., one in which it would gather and update a large amount of data from surveys and other sources on a continuing real-time basis), it could construct an integrated longitudinal dataset on the sample of science and engineering workforce members that could be queried for specific information at any time, and made similarly available to prospective analysts within and outside of NCSES. This would be a complex and challenging undertaking, but it would address many current problems entailed in getting descriptive data as of a given time to users in a timely fashion, while also facilitating analytical use of the data, particularly in full and longitudinal form.

CONCLUSION

In summary, there is an increasing need for NCSES's data collection and products to reflect nonlinear education and work paths, as well as a science and engineering workforce that is diverse in gender, race/ethnicity, and citizenship and is highly geographically mobile beyond the United States. One can imagine a future framework for NCSES's science and engineering workforce surveys that combines survey data and administrative and other nonsurvey data into a large and representative sample (or samples) of the workforce that is followed longitudinally in real time, using more specifically targeted survey modules to address a wider range of content and reducing the need for frequent fixed-interval follow-up surveys. In the future, it may be possible to draw samples increasingly and perhaps completely from institutional or other administrative records, including those capturing immigrants and emigrants. Ideally, science and engineering workforce data from all higher education levels could be linked, providing both descriptive and analytic users with fuller, timelier, and more accessible data on the entire spectrum of the workforce and the trajectories and pathways of its members over time and the life course.

Actualizing such an aspirational framework for NCSES would be challenging and require time and resources, but would represent movement toward an ultimately more integrated and efficient science and engineering workforce data system. Succeeding chapters further describe the current system and approach (Chapter 2) and present the panel's recommendations for both near-term improvements and the transition to a new framework in the medium or longer term.

2

Background on the Surveys

The mission of the National Center for Science and Engineering Statistics (NCSES) includes providing data on the science and engineering workforce, which NCSES defines as individuals educated or employed in a science and engineering field. Over the years, the science and engineering workforce surveys have provided valuable data to stakeholders, such as policy makers, the Office of Management and Budget, the Office of Science and Technology Policy, the National Science Board, other federal agencies, researchers, industry, the media, and the public. Based on online downloads of the National Science Board's *Science and Engineering Indicators (SEI)* report, data provided on the science and engineering workforce represent the most widely accessed statistics from the report, suggesting that these data are some of NCSES's most useful products (National Research Council, 2014).

THE NCSES HUMAN RESOURCES STATISTICS PROGRAM

The NCSES Human Resources Statistics Program provides data that help policy makers and researchers understand the continuum from precollege education through undergraduate enrollment, undergraduate degrees, graduate enrollment, graduate degrees, and the workforce. Several governmental agencies and organizations, including the National Center for Education Statistics (NCES) in the U.S. Department of Education (ED), and the U.S. Census Bureau, provide data on the different stages along this continuum. NCSES plays a major role in providing education data based on two censuses: (1) the Survey of Graduate Students and Postdoctorates in Science and Engineering (GSS), which provides annual data on current graduate enrollment in science, engineering, and selected health fields; and (2) the Survey of Earned Doctorates (SED), which provides annual data on research doctoral degrees received. NCSES provides workforce data based on the National Survey of College Graduates (NSCG), the Survey of Doctorate Recipients (SDR), and the Early Career Doctorates Survey (ECDS).

For each stage on the continuum from precollege education to participation in the workforce, the key information NCSES considers important to understand includes (Rivers, 2016) the following:

- How many people are at that stage?
- What are their demographic characteristics?
- What are the short-term and long-term trends at that stage?
- How does performance at one stage relate to performance at subsequent stages?

NCSES EDUCATION SURVEYS

GSS

Description

The GSS is an annual census survey of all academic institutions in the United States that award research-based master's degrees or doctorates in science, engineering, and selected health fields. The GSS has been conducted since 1972, and is sponsored by the National Science Foundation (NSF), the National Institutes of Health (NIH), and in some years the U.S. Department of Energy.

Through institutional coordinators, the GSS collects aggregate data on graduate students and postdoctoral appointees, as well as other doctorate-holding nonfaculty researchers, by demographic and other characteristics. These data enable the assessment of shifts in graduate enrollment and postdoctoral appointments and trends in financial support.

The GSS is completed at the level of organizational units (e.g., academic departments, degree-granting programs, university-affiliated research centers, and health care facilities), and as noted, the respondents are typically institutional coordinators. As of the most recent survey for which data are available (2015), there were 15,202 organizational units of interest at 711 academic institutions. The data collection is conducted in two stages: (1) verification of GSS-eligible units at the institution and (2) reporting of the data.

Institutional coordinators are initially contacted by e-mail and then followed up by telephone and FedEx letter. The overwhelming majority of institutional coordinators respond using a Web questionnaire (89.0%), but there is also an option to upload the data in a batch format (9.8%) or submit an Excel file (1.1%). The response rate for the 2015 GSS was 99.5 percent as a result of NCSES's extensive efforts to cultivate relationships with the institutions and the institutions' interest and willingness to be included in the study (see Chapter 5 for further discussion of the data collection methods). Box 2-1 shows the data collected as part of the GSS (see https://www.nsf.gov/statistics/srvygradpostdoc [January 2018] for the survey instrument).

NCSES has made several enhancements to the GSS over the past few years. In 2010, it expanded the postdoc section of the survey to collect detailed information on postdocs and doctorate-holding nonfaculty researchers (see Box 2-1). It also began requesting that a respondent knowledgeable about postdoc data be designated to complete the survey on behalf of the institution. In 2013, it added a checkbox to confirm that reported data excluded practitioner-oriented graduate degrees. In 2014, the survey frame was updated to add units from 151 newly eligible institutions and exclude 2 private for-profit institutions because they were offering primarily practitioner-based graduate degrees.

BOX 2-1
Main Topics in the Survey of Graduate Students and Postdoctorates in Science and Engineering

Graduate Students
- Degree field: science, engineering, health
- Enrollment status: full or part time, first time full time
- Demographics: sex, race/ethnicity, citizenship
- Major source of financial support: federal, nonfederal, self
- Major mechanism of financial support: fellowship, traineeship, research assistantship, teaching assistantship, other

Postdocs
- Demographics: sex, race/ethnicity, citizenship
- Major source of financial support: federal, nonfederal, self
- Major mechanism of support: fellowship, traineeship, research grants, other
- Type of doctorate: PhD, professional, dual
- Origin of doctorate: U.S., foreign

Doctorate-Holding Nonfaculty Researchers
- Demographics: sex
- Type of doctorate: PhD, professional, dual

SOURCE: Rivers (2016).

Some of the strengths of the GSS are that it provides long-term trend data on graduate student and postdoc counts in science, engineering, and selected health fields by citizenship, ethnicity, race, and sex. The survey also provides data on primary source of financial support for graduate students and postdocs by sex and field and on mechanism of financial support by field. The institution-level data can be used for peer institution comparisons. One of the limitations of the GSS is that it is becoming increasingly more difficult to distinguish research and practitioner-oriented degrees with the current data collection approach. Historically, another limitation was that the survey did not provide separate counts for master's or doctorate students because the data were available only at the aggregate level. However, NCSES plans to address this limitation with the 2017 survey cycle, when separate enrollment counts will be collected for master's and doctoral students.

Uses of the GSS Data

The GSS is the only national survey that provides data on graduate enrollment at the detailed field-of-study level by such institutional characteristics as highest degree granted, geographic location, public or private institution, medical schools, historically black colleges and universities, and land-grant institutions. Available information includes foreign enrollment, minority enrollment, field-specific enrollment, and so on.

One of the main uses of GSS data is to understand trends in the future stock of personnel trained in science, engineering, and selected health fields. The data are also used to analyze participation in selected fields by specific groups of individuals (such as minorities) or by types of institutions and to track student support patterns. GSS data are consulted to determine department eligibility for NSF programs targeted at special populations or types of programs and are used to assess the effects of NSF initiatives.

Data from the GSS are used by NSF to update the *SEI* and *Women, Minorities, and Persons with Disabilities in Science and Engineering* (*WMPD*) reports and are provided to Congress and the Executive Branch for a variety of uses. Data are also provided to NCES, which produces the *Digest of Education Statistics*. Tabulations are often prepared for NIH for use in program evaluation and to inform NIH budgets.

GSS data are also used by the academic institutions providing the data. NSF prepares the *Academic Institutional Profiles* by combining data from different sources, including the GSS, and addresses additional data requests from individual institutions and academic organizations. Academic institutions use the data to assess how they compare with other institutions (for example, in terms of federal support of their students or ability to reach special populations) and for planning purposes. Data from the GSS are also used in developing the Carnegie Classification of Institutions of Higher Education, a framework for classifying U.S. academic institutions for educational and research purposes.

Professional societies, such as the Association of American Universities, the Council of Graduate Schools, the American Society for Engineering Education, and the National Postdoctoral Association, use GSS data to track trends in graduate enrollment by fields of study. GSS data are also used by researchers from academia and other sectors to understand trends in graduate training.

NCSES has several ongoing activities aimed at increasing the usefulness of GSS data, including various outreach efforts to stakeholders. The agency held a data user workshop with federal and nonfederal stakeholders to better understand current uses of the data and increase the survey's usefulness. The agency also conducted an institutional coordinator survey to determine how data are maintained and accessed at different institutions and to assess institutions' potential ability to report master's and doctoral student data in more detail, such as separately by Classification of Instructional Programs codes. In addition, NCSES conducted a pilot survey during the 2016 survey cycle to determine the feasibility of collecting data separately for master's and doctoral students and postdoc data by file upload.

SED

Description

The SED is an annual census survey of all new recipients of research doctorates from research doctorate–granting institutions in the United States in a given academic year. A research doctorate is defined as a doctorate that requires an original contribution of knowledge to the field (typically in the form of a dissertation) and is not intended primarily for the practice of a profession (for example, as an M.D. or J.D. would be). Doctorate recipients in all fields are included, not just those in science and engineering.

The SED has been conducted since academic year 1958 and is sponsored by NSF, NIH, ED, the U.S. Department of Agriculture (USDA), the National Endowment for the Humanities (NEH), and the National Aeronautics and Space Administration (NASA). Data from the survey provide an assessment of characteristics of the doctoral population and of trends in doctoral education and degrees.

In 2015, the universe for the SED included 432 academic institutions and approximately 55,000 doctorates. Institutional coordinators facilitate the distribution of information about the survey around the time of degree completion, and NCSES's data collection contractor follows up with nonrespondents using e-mail, letters, postcards, and telephone calls (for a more detailed discussion of the data collection methods, see Chapter 5). Most questionnaires are completed by Web (93.0%), but mail (5.4%) and telephone (1.6%) modes are also available. The overall response rate is around 90 percent.

Box 2-2 shows the data collected as part of the SED. The full questionnaire can be viewed at https://www.nsf.gov/statistics/srvydoctorates/ [January 2018] and a crosswalk of the survey questions is included in Appendix B.

NCSES has made several changes to the SED over the years to increase the survey's efficiency and usefulness. Between 2010 and 2015, reliance on the Web as a primary mode for the survey increased substantially. In 2011, disability-related questions were replaced with questions about functional limitations. In 2014, the means of collecting salary data in the Web survey mode was changed to an approach that splits the item over two screens, resulting in a nearly 30 percent increase in reporting of exact salaries instead of salary ranges. In the same year, new items were added to enhance education timeline analyses. In 2015, additional improvements were made to the Web survey with the addition of a field-of-study search function and optimization for mobile devices.

BOX 2-2
Main Topics in the Survey of Earned Doctorates

- Degree history (type, year, field, institution)
- Financial support during graduate school
- Education-related debt incurred
- Time to degree
- Employment plans for the coming year (1 year after degree completion)
  - Postgraduation status (e.g., definite commitment, searching for position)
  - Type of position (postdoc vs. employment)
  - Employment sector
  - Postgraduation location
  - Primary/secondary work activity
  - Salary
- Demographics
  - Sex
  - Race/ethnicity
  - Marital status
  - Citizenship status
  - Disability status

SOURCE: Rivers (2016).

Among the strengths of the SED are that it provides reliable counts of research doctorates earned in the United States by demographic and degree characteristics. The survey also provides information on postgraduation plans for employment, including lack of firm plans (reported by 38% of respondents in the 2015 SED). Data are available for research doctorates as far back as 1920. The survey also serves as a sampling frame for the SDR.

Although the SED manages to obtain very high response rates, maintaining these response rates is increasingly difficult for both institutions and individuals. Among other challenges mentioned by NCSES is researchers' growing interest in linking SED data to administrative records maintained by the institutions, which raises confidentiality concerns. Researchers are also interested in access to a public dataset and online analysis tools.

Uses of the SED Data

The SED is the only source of national data on the education and career plans of research doctorate recipients from U.S. academic institutions. This data collection complements the GSS, which collects data on the training of graduate degree recipients. Federal agencies use the data for program evaluation and planning purposes. The White House Office of Science and Technology Policy and the Congress also use the data to inform policy discussions and the development of legislation on such topics as doctorates awarded to temporary visa holders and student debt levels. The six federal agencies that sponsor the survey—NSF, NIH, ED, USDA, NEH, and NASA—are particularly heavy users of the data.

NSF uses the SED data in a series of internal and external reports, including the *SEI* and *WMPD* reports. Several NSF programs focused on women, minorities, and persons with disabilities, as well as fellowship programs, use the data for evaluation and planning purposes. In addition, SED data have informed the work of NSF's Directorate for Education and Human Resources and of an interagency committee on foreign access to U.S. science and engineering at American academic institutions. As part of its participation in the international Careers of Doctorate Holders project, NCSES has also provided SED data to the OECD. The SED data serve an important role as well as the sampling frame for the SDR.

NIH has used SED data to monitor the volume of doctorate recipients in the biomedical, behavioral, and clinical sciences. This information feeds into the agency's efforts to evaluate and plan its research training programs and grants.

ED, particularly NCES, uses SED data in its annual *Digest of Education Statistics* and for occasional other reports. SED data are also used to evaluate several ED programs, including those of the Office of Federal Student Aid and Office of Planning, Evaluation, and Policy Development.

USDA uses SED data to monitor trends in fields of particular interest to the agency. The data are also used for evaluation and planning purposes for several Land Grant and Tribal College programs. In addition, USDA publishes tables based in part on SED data on the Food and Agricultural Education Information System Website.

NEH and NASA use SED data to monitor trends in the number and characteristics of doctorate recipients in the fields that are most relevant to these agencies' missions.

SED data are also used for evaluation and planning purposes by the academic institutions that provide the data. Detailed data are provided to graduate school deans on the characteristics of their degree recipients, along with comparisons with national data and data for peer institutions. Institutions use the data to analyze such questions as minority enrollment or the time it takes to obtain a degree.

Finally, SED data are used by researchers from other sectors, including academic researchers.

NCSES WORKFORCE SURVEYS

Figure 2-1 shows the population coverage for the three main NCSES surveys that are focused primarily on the workforce. The NSCG is the most comprehensive in terms of population coverage, including individuals with degrees in both science and engineering fields, as well as other fields, and regardless of type of terminal degree or when and where the degree was awarded. The SDR focuses on individuals with research doctoral degrees in science, engineering, or selected health fields obtained from academic institutions in the United States. The ECDS sample includes individuals with doctoral degrees from abroad and individuals with a degree in a nonscience and engineering field, but unlike the SDR, focuses on recent graduates only.

NSCG

Description

The NSCG has been conducted approximately biennially since 1972. The sample includes individuals with a bachelor's degree or higher who are under the age of 76 and reside in the United States in a noninstitutionalized setting, regardless of whether the person obtained a degree in the United States or abroad. The survey provides data on college graduates but has a particular focus on those in the science and engineering workforce.

The current NSCG uses a rotating panel sample design. Each new sample is selected from the American Community Survey (ACS), a large

Educational Attainment	Occupation Type	Degree Field	Degree Type	Degree Date	Population Coverage		
					NSCG	**SDR**	**ECDS**
Bachelor's degree or higher	S&E or S&E-related Occupation	S&E or S&E-related Degree Field	U.S. BS/MS	N/A	NSCG		
			U.S. Doctorate	10 years or less		SDR	ECDS
				> 10 years			
			Foreign BS/MS	N/A			
			Foreign Doctorate	10 years or less			ECDS
				> 10 years			
		Non-S&E Degree	U.S. BS/MS	N/A			
			U.S. Doctorate	10 years or less			ECDS
				> 10 years			
			Foreign BS/MS	N/A			
			Foreign Doctorate	10 years or less			ECDS
				> 10 years			
	Non-S&E Occupation	S&E or S&E-related Degree Field	U.S. BS/MS	N/A			
			U.S. Doctorate	10 years or less		SDR	ECDS
				> 10 years			
			Foreign BS/MS	N/A			
			Foreign Doctorate	10 years or less			ECDS
				> 10 years			
		Non-S&E Degree	U.S. BS/MS	N/A			
			U.S. Doctorate	10 years or less			ECDS
				> 10 years			
			Foreign BS/MS	N/A			
			Foreign Doctorate	10 years or less			ECDS
				> 10 years			

FIGURE 2-1 Population coverage for the National Survey of College Graduates, Survey of Doctorate Recipients, and Early Career Doctorates Survey.
SOURCE: Finamore (2016).

ongoing survey that is conducted by the U.S. Census Bureau and uses monthly samples to produce estimates for small areas, such as census tracts and block groups. The ACS includes questions on degree level, bachelor's degree field of study, occupation, and demographic characteristics, enabling NCSES to select a sample for the NSCG without having to conduct a separate costly screening of households. Sample members are interviewed every 2–3 years for four cycles before rotating off the panel. For the 2015 data collection, the overall sample size was approximately 135,000, including old and new cases, and the response rate was 70 percent.

NCSES uses several modes of contact to communicate with NSCG sample members, including letter, postcard, e-mail, and telephone; sample members can respond by Web (77%), mail (15%), and telephone (8%). The data collection process is discussed in further detail in Chapter 5.

Box 2-3 shows the NSCG data collected. The full questionnaire can be viewed at https://www.nsf.gov/statistics/srvygrads [January 2018].

In addition to its core items, the survey occasionally also includes special modules. Modules over the course of the years have included the following:

- Professional certifications: 2015
- Community college enrollment: 2013, 2015
- Education debt (amount borrowed/owed): 2013, 2015
- Job satisfaction and attributes: 2003, 2010, 2013, 2015
- Federal support of work: 2003, 2010, 2013, 2015
- Immigration information: 2003, 2010, 2013, 2015
- Organization of work: 2006
- International collaboration: 2006
- Academic positions: 2003
- Productivity measures (publications, patenting): 2003

Since the 2010 implementation of the rotating panel design, recent changes to the NSCG have included the implementation of a young graduates

BOX 2-3
Main Topics in the National Survey of College Graduates

- Employment
  - Labor force status
  - Characteristics of job and employer
  - Work activities
  - Relationship of education and occupation
- Education
  - Complete degree history (type, year, field, institution)
  - Community college enrollment
  - Financial assistance for education
  - Continuing education activities
- Demographics
  - Sex
  - Race/ethnicity
  - Marital status
  - Citizenship status
  - Disability status
  - Number of children

SOURCE: Finamore (2016).

oversample in 2013, designed to replace the discontinued National Survey of Recent College Graduates (NSRCG)[1] as a source of information on recent graduates. In the same year, adaptive design was implemented, and emphasis on the Web as the preferred mode of response was increased to improve the efficiency of the data collection.

In addition to reliable estimation of the science and engineering workforce by demographic, occupation, and degree characteristics, the strengths of the NSCG include estimation of foreign-earned degree holders and the ability to measure the inflow of new graduates into the science and engineering workforce. In addition, the data enable researchers to compare estimates of the science and engineering population with estimates of other groups. In terms of challenges, despite the survey's relatively high response rates, the NSCG is struggling with increasing nonresponse, as is the case with other surveys. The data available from the NSCG also are not as detailed as some researchers would like—for example, in the areas of geography, race and ethnicity, and occupation. In addition, the survey's dependence on the ACS for its sampling frame represents a unique challenge because that survey is overseen by another statistical agency, the U.S. Census Bureau, and faces complex challenges of its own.

Uses of the NSCG Data

The NSCG is the most comprehensive source of national data on college graduate scientists and engineers in the U.S. workforce. The data are used by policy makers; federal, state, and local government agencies; industry; professional organizations; educational institutions; researchers; and others. Frequently examined questions include the impact of foreign-born and foreign-educated scientists and engineers in the U.S. workforce, characteristics of the transition from higher education to the workforce, issues related to diversity, and the implications of an aging cohort of scientists and engineers. NSF regularly produces reports examining workforce issues using the NSCG data. The National Science Board uses NSCG data in the *SEI* reports, and has also used the data in occasional additional reports focused on national workforce policies. NCSES uses NSCG data in the *WMPD* reports.

Another regular federal use of the NSCG data is a biennial congressionally mandated report by the Committee for Equal Opportunities

[1] The NSRCG was a survey conducted by NCSES between 1973 and 2010. The survey sampled individuals who received a bachelor's or master's degree in a science, engineering, or health field from a U.S. academic institution in the 2 or 3 academic years prior to each biennial data collection. The NSRCG provided data on the education and employment characteristics of recent college graduates, and it also served as a primary source of recent college graduates for the NSCG sample. For a more detailed discussion of the NSRCG, see Chapter 4.

in Science and Engineering (CEOSE) highlighting areas of concern related to equal opportunities for participation in science, technology, engineering, and mathematics (STEM) fields. In the past, the Executive Office of the President also has used NSCG data to understand the contribution of immigrants in science and engineering occupations.

Finally, the NSCG is used by nonprofit education organizations and associations. For example, the Council of Graduate Schools (CGS) uses NSCG data to estimate the return on investment for obtaining an advanced degree. In collaboration with the Educational Testing Service, CGS also has used NSCG data to examine national benchmarks for career outcomes of graduate degree recipients by field. The data are used by academic researchers as well.

SDR

Description

The SDR is a longitudinal survey conducted biennially since 1973. It provides demographic and career history information about individuals with a U.S. research doctoral degree in a science, engineering, or selected health field who are under the age of 76 and in a noninstitutionalized setting. Beginning with the 2015 survey cycle, those surveyed included U.S.-trained research doctoral degree recipients living abroad.

The SDR uses the SED as the sampling frame and provides cross-sectional trend data on the employment and occupation of degree recipients since their degree was awarded. The data used in the design of the stratified sample include race/ethnicity, sex, age, disability status, degree field, postgraduation plans for staying in or leaving the United States, year of degree, citizenship status, and place of birth. The sample design is a fixed panel, with each survey cycle consisting of the prior round's sample along with a new cohort of the most recent graduates. The sample size in 2015 was approximately 120,000 cases.

Most sample members are contacted to participate in the SDR for the first time approximately 2–3 years after receiving their doctoral degrees, based primarily on contact information obtained at graduation. For this reason, and because this survey does not benefit from the assistance of institutional coordinators as does the SED (sample members are contacted directly by NCSES's data collection contractor), the response rate is much lower than that of the SED. Respondents can participate by Web (82%), mail (10%), and telephone (8%); the overall response rate for the most recent survey was 63 percent.

BOX 2-4
Main Topics in the Survey of Doctorate Recipients

- Employment
  - Labor force status
  - Characteristics of job and employer
  - Work activities
  - Relationship of education and occupation
  - Academic position/faculty rank
  - Tenure status
- Education
  - Complete degree history (type, year, field, institution)
  - Continuing education activities
- Demographics
  - Sex
  - Race/ethnicity
  - Marital status
  - Citizenship status
  - Disability status
  - Number of children

SOURCE: Finamore (2016).

Box 2-4 shows the data collected as part of the SDR. The full questionnaire can be viewed at https://www.nsf.gov/statistics/srvydoctoratework [January 2018].

In addition to the core questions, the SDR has included modules on the following special topics over the years:

- Job satisfaction and attributes: 2001, 2003, 2010, 2013, 2015
- Federal support of work: 2003, 2010, 2013, 2015
- Immigration information: 2003, 2010, 2013, 2015
- Organization of work: 2006
- International collaboration: 2006
- Productivity measures (publications, patenting): 1995, 2001, 2003, 2008
- Postdoctoral experiences: 1995, 2006

A number of enhancements have been made to the SDR in recent years. First, like the NSCG, the SDR recently implemented adaptive design strategies to increase the efficiency of the data collection, reduce nonresponse bias, and maximize the representativeness of several key domains.

In addition, the SDR sample has undergone several redesigns over the years. Most notably, in 2015 the sample was substantially increased from 40,000 to 120,000. The expansion included a cohort of doctorate recipients with degrees awarded since 1961, a cohort of new doctorate recipients with degrees awarded in 2012 and 2013, and a subsample of the 2013 SDR panel. The purpose of the expansion was to improve estimation of employment outcomes at the fine field-of-degree level (at the level of distinct subfields, as opposed to only broad fields) and improve the existing aggregated degree-level estimates for domains defined by various demographic characteristics. The increased sample size is expected to improve researchers' ability to conduct analyses of small populations, including by gender, ethnicity, and race.

Another strength of the redesigned SDR is that it can track international mobility by following up with sample members who have moved abroad (the international follow-up was piloted in the 2003, 2006, and 2008 survey cycles and referred to as the International Survey of Doctorate Recipients). However, the survey still does not allow reliable estimation across all 215 fine fields by demographic characteristics because of small sample sizes in some fields. Increasing nonresponse also is a challenge for the SDR, especially within the expansion cohort. Another challenge for the survey is locating respondents who have been in the sample for a long time, especially those who have moved abroad, and the proportion of sample members living abroad has been increasing. The recent sample redesign introduced an additional locating challenge because a new cohort of doctorate recipients with degrees awarded since 1961 was added to the sample for the first time in 2015. For these sample members, the only contact information available was that current at the time the degree was awarded. These factors contributed to a predictable drop in the SDR response rate, from 76 percent in 2013 to 63 percent in 2015 (by comparison, the response rate for the NSCG declined from 74 percent in 2013 to 70 percent in 2015). The issue with the SDR will be slightly ameliorated as new graduates are added to the sample in each cycle, and NCSES is working with its data collection contractor to refine the locating procedures.

Uses of the SDR Data

The SDR provides the most comprehensive data on scientists and engineers with research doctoral degrees from U.S. institutions. The data are used by policy makers; federal, state, and local government agencies; industry; professional organizations; educational institutions; researchers; and others. Issues examined using SDR data include the impact of foreign-born U.S.-trained scientists and engineers in the U.S. workforce, the role of postdocs in careers, characteristics of the transition from higher education to the workforce, issues related to diversity, and the implications of an aging cohort of scientists and engineers.

NSF produces a variety of reports using the SDR data. As is the case with the data from other NCSES workforce surveys, the National Science Board is one of the regular users of the SDR for its *SEI* and companion reports. NCSES also uses SDR data in the *WMPD* reports. In addition, the NSF Education and Human Resources Directorate frequently uses SDR data for the development and evaluation of programs. Aside from NSF, NIH, USDA, NASA, the Department of Energy, and the Department of Commerce use SDR data for a variety of planning purposes. And CEOSE uses SDR data in its biennial congressionally mandated report on equal opportunities for participation in STEM fields.

NCSES also has provided SDR data to OECD. In addition, the data are used by academic researchers examining questions related to the U.S. workforce.

ECDS

Description

The ECDS is a new cross-sectional NCSES survey designed to provide estimates of the work-related experiences of early career doctorates who earned their first doctorate or doctorate-equivalent degree within the past 10 years and are working at an academic institution (as defined in the GSS), at a federally funded research and development center (FFRDC), or in the NIH Intramural Research Program (NIH IRP). A pilot of this survey was conducted in 2014–2015, and the first full-scale data collection has begun. Although the sampling frame is limited in terms of coverage, at least at this stage in the survey's development, the ECDS is intended to fill an important gap in understanding of the circumstances and characteristics of early career doctorates.

The ECDS uses a two-stage sample design, which includes a stratified probability sample of institutions followed by a stratified probability sample of individuals. The pilot study included a sample of 176 institutions and 6,872 individuals. The next round of the survey is expected to include 350 institutions and 22,855 individuals.

Almost all of the pilot study interviews (99.5%) were conducted by Web and the remainder by telephone. The overall response rate was 56 percent, with 84 percent in the first stage, and 66 percent in the second stage.

Box 2-5 shows the topics covered in the ECDS. The full questionnaire can be viewed at https://www.nsf.gov/statistics/srvyecd/ [January 2018]. The ECDS is the first survey to provide national estimates of the early career doctorate population working at GSS academic institutions, FFRDCs, and

BOX 2-5
Main Topics in the Early Career Doctorates Survey

- Education
  - Degrees earned/type
  - Primary field of study
  - Origin of degrees (U.S. vs. non-U.S.)
  - Education-related debts
- Employment
  - First position after doctorate
  - Current position
  - Work activities
  - Compensation and benefits
  - Interactions with managers
  - Relationship of education and occupation
- Demographics
  - Sex
  - Race/ethnicity
  - Marital status
  - Citizenship status
  - Disability status
  - Number of dependents
- Additional Topics
  - Professional activities and achievements
  - Professional and personal life balance
  - Mentoring, training, and research opportunities
  - Career development and plans

SOURCE: Finamore (2016).

the NIH IRP. The survey determines common definitions for the different types of early career doctorates that can be applied across and within sectors. Other strengths of the survey are that it allows for comparisons between estimates of scientists and engineers and estimates of other groups, as well as for comparisons between U.S. and non-U.S. doctoral degree holders. As with other surveys, nonresponse is a challenge, and the pilot study experienced particular difficulty with obtaining buy-in from institutions. Other challenges include the difficulty of obtaining a comprehensive sampling frame for employment sectors beyond the GSS academic institutions, FFRDCs, and the NIH IRP, and the lack of demographic data available for use in the stage 2 sampling design.

Uses of the ECDS Data

NCSES released data from the ECDS pilot study in the 2017 *WPMD* report and in an InfoBrief. The ECDS is a unique source of detailed information on early career doctorates, including their employment and productivity, career pathways followed, and future career plans. Beyond being able to expand the data in the *SEI* reports and produce additional reports focused on this population, NCSES anticipates that the ECDS data will be useful to government agencies, such as NIH, and to organizations, such as CGS, the Association of American Universities, the American Association of Medical Colleges, the National Postdoctoral Association, and the American Association for the Advancement of Science. NCSES also expects to share the data with OECD.

Survey of Postdocs at FFRDCs

In addition to the workforce surveys described above, since 2005 NCSES has conducted the Survey of Postdocs at FFRDCs every 2–3 years. The survey collects aggregate data on postdocs by citizenship, ethnicity, race, and sex, as well as field of research and citizenship. The frame for this survey includes all FFRDCs in the Master Government List maintained by NSF and the NIH IRP. The survey is conducted by Web, and typically achieves a 100 percent response rate.

PRIOR COMMITTEE ON NATIONAL STATISTICS STUDIES

Over the years, the National Academies of Sciences, Engineering, and Medicine's Committee on National Statistics (CNSTAT) has convened several panels to assist NCSES (formerly the Division of Science Resource Statistics) with issues pertinent to the agency's mission. Some of these reports have played a particularly important role in guiding NCSES's approach to measuring the science and engineering workforce (National Research Council, 1989, 2003, 2008).

One of the key reports, issued at the end of the 1980s, focused on the NSF surveys that made up what was at the time called the Scientific and Technical Personnel Data System (National Research Council, 1989). The database included data from the NSF Postcensal Survey, which was a survey of individuals in science and engineering fields sampled based on the decennial census, and reinterviews with the same respondents as part of the NSF New Entrants Survey. The data system also included the SED and SDR. The CNSTAT panel recommended a redesign of the data system, in part to address reliability issues. The ensuing design changes led to the establishment of the Scientists and Engineers Statistical Data System

(SESTAT). The panel that authored a second report revisited the SESTAT design and recommended modifications to prepare SESTAT for the beginning of the new century (National Research Council, 2003).

A few years later, a CNSTAT panel examined options for addressing a challenge faced by NSF as a result of the elimination of the decennial census long form, which had served as the source of information on college graduates for the sampling frame of the survey that came to be known as the NSCG (National Research Council, 2008). The panel recommended the use of the Census Bureau's new ACS as a sampling frame for the NSCG. The panel's input led to the implementation of a rotating panel design for the NSCG and the discontinuation of the NSRCG. This report also set the stage for further evaluation of NSF's approach to measuring the science and engineering workforce.

A recent CNSTAT panel examined the status of science, technology, and innovation indicators being published by NCSES and whether new or revised indicators are needed (National Research Council, 2014). Some of the relevant recommendations from that panel's report are that NCSES

- consider options for exploiting the longitudinal panel structure of the SDR to create indicators of researcher mobility over time and track the initial employment and career paths of recent doctorate recipients in the first years after they receive their doctoral degrees;
- draw on occupation data from the Longitudinal Employer-Household Dynamics Program and education data from the Baccalaureate and Beyond Longitudinal Study to create indicators of labor mobility, particularly for industries with high growth and those in which the United States has a competitive advantage;
- enhance data on STEM subpopulations, such as early career doctorate recipients, master's degree holders, and community college graduates;
- explore the possibility of utilizing full-text resources of dissertation databases to create new indicators, for example, on the relatedness of different fields; and
- consider using ACS data to produce indicators that track salaries in different occupations and for college graduates receiving different types of degrees.

NCSES has begun addressing some of these recommendations, as well as making various improvements noted in this report. The recommendations in Chapters 3–6 reinforce some of those made by the 2014 CNSTAT panel and address other directions NCSES is taking, while raising further issues relevant to the present panel's charge.

3

Evolving Data Needs

The rapidly changing characteristics of the science and engineering workforce result in a concomitant change in data needs. The National Center for Science and Engineering Statistics (NCSES) has always been focused on understanding these needs and keeping its surveys current. The recently launched Early Career Doctorates Survey (ECDS) is an example of an initiative that is responsive to emerging issues and to recommendations made by previous panels of the National Academies of Sciences, Engineering, and Medicine as well.

Part of NCSES's motivation for commissioning the present study was to ensure that the agency is continuing to meet the policy and research needs of stakeholders, including Congress, federal agencies, the Office of Management and Budget, the National Science Board, researchers in academia and other sectors, providers of the data, the media, and the public. The challenges posed by the competing needs of stakeholders for surveys of this type are substantial, primarily because of the need to limit respondent burden. In addition, the need to keep pace with new developments in the realm of the science and engineering workforce has to be balanced with the need to maintain stability in the estimation of key trends.

Potential increases in respondent burden resulting from efforts to meet stakeholders' data needs are a major concern because they could have an adverse impact on response rates, and therefore on data quality. Ideally, if new questions were added, other questions that had become obsolete or were not widely used would be dropped, and other questions might be asked of more targeted segments of the sample or in response to specific information gathered from survey or monitoring data. However, determining what questions to drop or ask in more targeted or selective ways can be a difficult task for any government survey with a wide range of stakeholders who depend on the data, and who may or may not be active in making their voices heard. Identifying questions that could potentially be dropped is beyond the scope of this study. However, information about what data and publications are downloaded, as well as input gathered through discussions with data users at various venues, can provide insight into whether some items are no longer used. At the same time, these methods need to be used with caution to ensure that they do not result in misleading conclusions. For example, users of the data tables may be very different from users who typically download the microdata.

WORKSHOP ON DATA NEEDS

To better understand the data needs of users of science and engineering workforce data, including topics that are increasing in importance as a result of changes in society and the workforce, the panel held a workshop on October 20, 2016 (the workshop agenda, including the list of speakers, appears in Appendix C). The workshop involved a diverse group of participants from a variety of backgrounds and disciplines. Participants emphasized that NCSES enjoys international recognition for its work in the area of science and engineering statistics, and that the data it produces are of very high quality and meet important policy and research needs. The congressionally mandated *Science and Engineering Indicators* (*SEI*) and *Women, Minorities, and Persons with Disabilities in Science and Engineering* reports produced by NCSES were among the data products highlighted as most widely used. The new international component of the Survey of Doctorate Recipients (SDR)—the International Survey of Doctorate Recipients from U.S. institutions who migrate abroad—was mentioned as a recent addition that addresses an increasing need for data on such individuals.

Workshop participants shared their views on existing and emerging data needs, and several of the key areas discussed are reviewed in this chapter. Many of these needs go beyond NCSES's primary mandate and should not be viewed as shortcomings of the existing surveys, but as areas in which there is potential for enhancing the usefulness of the data. Furthermore, the input received from stakeholders indicates that the content of the surveys matches many data needs reasonably well, yet the data are not always available in ways that would maximize analytic capabilities, particularly longitudinal analysis, and other emerging areas of interest are just beginning to be represented in the surveys. The nature of the data needs identified at the workshop suggests several possible mechanisms that could enable NCSES to accommodate new analyses or additional detailed questions about some topics or produce new estimates linked to the surveys in other ways, thereby responding to changes in education and the workforce, including demographic characteristics and highly flexible, nonlinear education and career paths.

This chapter reviews the key data needs going forward based on input received at the workshop and from other sources, including the panel members' own expertise and deliberations, as well as the broader literature on the science and engineering workforce. Some of the discussion in this chapter refers to NCSES's science and engineering workforce surveys in general, while some is specific to one survey. It is also important to acknowledge that "scientists and engineers" is a broad concept, and there are many differences among individual fields. Some of the comments made by data users apply to some fields but not others. It is beyond the scope of this report to address these differences, and indeed the nature of the surveys is such that it is not possible to collect detailed data specific to particular subfields.

It is important to note as well that some aspects of data users' needs are discussed in other chapters. For example, as a result of rapid changes in science and engineering education and workforce needs, a theme that emerged from the workshop is the desire for more current estimates across several topics. This issue is discussed in further detail in Chapter 6. Limitations related to data access, such as the level of detail that can be released, also are discussed in Chapter 6.

The remainder of this chapter examines possible strategies for addressing the identified data needs in ways that are likely to bring the most benefit with minimal disruption to trend data and the least increase in respondent burden. Expanding the data available through NCSES will undoubtedly necessitate reliance on creative data collection methods, such as administering topic modules, or asking some questions at longer intervals (depending on the goals for the question) or only if triggered by some event, such as a job change, as well as the increased use of nonsurvey and nontraditional sources of such data.

CAREER PATHWAYS

From both scientific and policy perspectives, it is increasingly imperative to understand the career pathways of scientists and engineers. NCSES recognizes, as stressed by workshop participants as well, that the education and career movements of scientists and engineers no longer follow a typical linear pipeline, but now are characterized by complex flows and pathways (Freeman and Goroff, 2009; Martin and Pearson, 2005; Morrison et al., 2011; Stephan, 2009, 2011). Movement increasingly occurs between public- and private-sector jobs and between academic and nonacademic jobs (Musselin, 2010). Some students take longer to graduate because they work part time or take time off for nonacademic training, such as internships. In some engineering and computer science fields, there is a perception that entering the workforce after obtaining a bachelor's degree pays off more than pursuing a doctorate (Ehrenberg and Kuh, 2009). Some complete a doctorate but work in a field other than that of their degree (Aanerud et al., 2007; Nerad, 2009) or shift fields as a result of interdisciplinary work. In some disciplines, several postdoctoral positions are not unusual, while in others such positions in general are rare (Nerad and Cerny, 2002). Some individuals end up in science and engineering careers by following nontraditional pathways, and the likelihood of this occurring varies by such characteristics as race and gender. Likewise, retirement may be followed by a second or third career. Figure 3-1 illustrates these pathways in the case of doctorate recipients in the biomedical sciences.

While the NCSES surveys capture at each point in time the different types of jobs held by individuals educated or employed in a science and engineering field, the trajectories involved are not well understood. In addition to understanding these pathways, policy makers and researchers would like to know more about what factors influence people's options and preferences. Some of these options and preferences evolve during academic training (Gibbs et al., 2014), and graduate deans are interested in the types of factors that influence people's initial career decisions, particularly decisions to seek employment in nonacademic sectors (Goulden et al., 2009). Other factors come into play at later points in the life course and career (see the sections on work environments, work–life balance, and discrimination and harassment below).

There is substantial interest as well in better understanding the role of postdoctoral positions in a career. It is unclear to what extent pursuing such a position and the type of position pursued (for example, academic or nonacademic) reflect career preferences, a lack of other options, a lack of clarity as to opportunities, or family-related decisions (Nerad and Cerny, 1999; Rudd et al., 2010). A question is whether a postdoctoral position should be considered primarily training, and ultimately what the impact of such positions is on subsequent careers. The recently launched ECDS is a welcomed attempt on the part of NCSES to address some of the key questions related to this population, but the sampling frame for that survey has more severe limitations relative to the other NCSES science and engineering workforce surveys (see Chapter 4). As discussed below, the panel encourages NCSES to consider merging the content and samples of the ECDS with the ongoing SDR.

The rapid changes occurring in the science and engineering workforce, particularly the evolving skills required, mean that in some fields there is a need for periodic retooling, which may result in a return to an educational setting after several years in the workforce or a change in careers. The turnover associated with this is not well captured by the surveys. Also needed

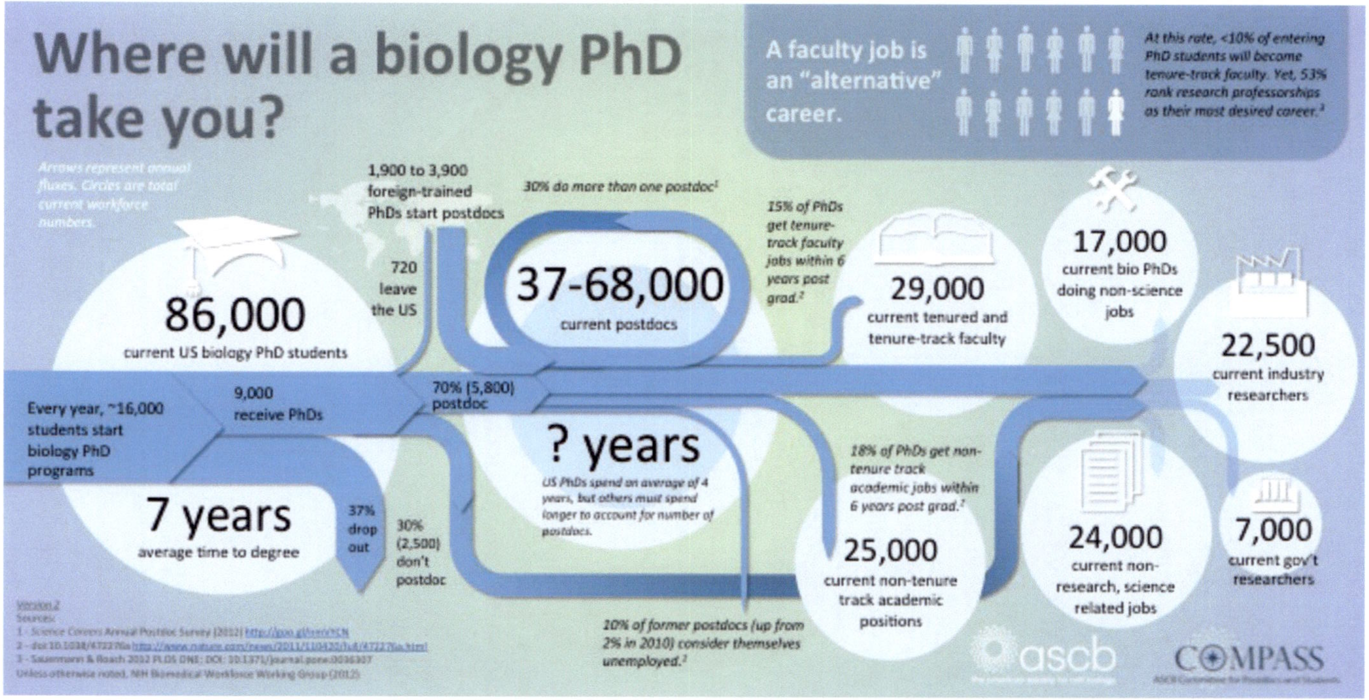

FIGURE 3-1 Current workforce sizes and annual fluxes before and after receiving a doctorate in the biomedical sciences.
SOURCE: Used with permission from Polka (2014) and ASCB, the American Society for Cell Biology.

is a better understanding of underemployment, the discouraged worker category, and re-entry in the context of science and engineering fields.

RECOMMENDATION 3-1: The National Center for Science and Engineering Statistics should place greater emphasis on following sample members longitudinally and obtaining survey data on the nature, determinants, and consequences of significant transitions in science and engineering career pathways.

INTERNATIONAL MOBILITY AND COMPARISONS

The increase in international mobility greatly complicates the career trajectories of many scientists and engineers. From a research and policy perspective, it is important to understand the mobility of scientists and engineers who are foreign-born (whether they are educated in the United States or not), as well as those who are born in the United States but are educated abroad, work abroad, or both (Lin et al., 2009). Of particular interest are the stay rates for foreign students and their characteristics in terms of field of degree and type of job (Gupta et al., 2003).

As discussed in Chapter 2, the SDR recently expanded its coverage to include individuals who have earned a doctorate from an institution in the United States even if they moved abroad after graduating. While collecting data from individuals who live abroad represents a challenge from a survey operations perspective, and in particular in terms of response rates, it is now possible to conduct analyses of this population based on whether they were U.S. citizens or temporary visa holders at the time they graduated, as well as on whether their place of residence was in the United States or abroad when the survey was conducted. Researchers at Oak Ridge Associated Universities have been producing regular estimates of the stay rates of foreign recipients of U.S. doctorates since 2000 (Finn, 2010, 2012, 2014). Over the years they have relied on data from the Social Security Administration (SSA) to produce these estimates, but because of difficulties associated with obtaining the data from SSA, their most recent report relies more heavily on the SDR (Finn, 2014).

The National Survey of College Graduates (NSCG) collects data on the college-educated population residing in the United States at the time of the survey, including individuals who received their degrees from abroad. However, the data available on the NSCG sample are more limited than those collected through the Survey of Earned Doctorates (SED) and SDR, and respondents are not reinterviewed if they move abroad (see Chapter 4).

Increasingly, members of the U.S. science and engineering workforce are people who have migrated from other countries to the United States after or in the midst of their education. NCSES has been working to identify such individuals and follow them as it does scientists and engineers who are educated and/or begin their careers in the United States. NCSES needs to maintain and enhance its efforts in these directions.

In addition to more comprehensive coverage of scientists and engineers who are mobile across countries and the ability to examine some of the same questions about career pathways discussed above, policy makers and researchers would like to better understand trends that are unique to the internationally mobile population. People who immigrate or emigrate within multinational organizations are of particular interest. One of the main questions is related to the factors that influence a person's decision about whether to remain in the United States or move abroad. This question is of interest not only nationally but also often from a local perspective. State and local governments frequently want to know what proportion of foreign students are remaining in the area where they graduate and why, especially in light of incentives by some countries (for example, China and India) to attract back doctorate recipients who studied abroad.

RECOMMENDATION 3-2: The National Center for Science and Engineering Statistics should work toward survey design and questionnaire content that capture both the immigration of individuals from abroad into the U.S. science and engineering workforce and the emigration of individuals from the U.S. science and engineering workforce to other countries, including the nature, determinants, and employment outcomes of such movements. Specific questionnaire modules targeted to these groups could explore such issues.

WORK ACTIVITIES, TRAINING, AND SKILLS

The NCSES surveys use a yes/no format to ask about the work activities in which respondents are engaged at least 10 percent of the time during a typical work week, and these questions provide a very basic understanding of the type of work people do. Figure 3-2 shows the NSCG and SDR list of work activities, which includes 13 items, such as basic research, applied research, and development. The question on work activities in the ECDS pilot survey, shown in Figure 3-3, includes a list of 10 items customized for early career doctorate recipients working in the types of settings covered by the sampling frame (academic institutions, federally funded research and development centers [FFRDCs], and the National Institutes of Health's Intramural Research Program). Because of the ECDS's more narrowly defined target population, its items are better able to reflect the work activities that are characteristic of the population of interest.

The NSCG and SDR also ask about work-related training, certification, and licenses, but less is known about the skills that respondents would

The next question is about your work activities on your principal job. Which of the following work activities occupied at least 10 percent of your time during a typical work week on this job?

Mark Yes or No for each item.

		Yes	No
1	Accounting, finance, contracts	1☐	2☐
2	Basic research – study directed toward gaining scientific knowledge primarily for its own sake	1☐	2☐
3	Applied research – study directed toward gaining scientific knowledge to meet a recognized need	1☐	2☐
4	Development – using knowledge gained from research for the production of materials, devices	1☐	2☐
5	Design of equipment, processes, structures, models	1☐	2☐
6	Computer programming, systems or applications development	1☐	2☐
7	Human resources – including recruiting, personnel development, training	1☐	2☐
8	Managing or supervising people or projects	1☐	2☐
9	Production, operations, maintenance (e.g., chip production, operating lab equipment)	1☐	2☐
10	Professional services (e.g., health care, counseling, financial services, legal services)	1☐	2☐
11	Sales, purchasing, marketing, customer service, public relations	1☐	2☐
12	Quality or productivity management	1☐	2☐
13	Teaching	1☐	2☐
14	Other – *Specify*	1☐	2☐

FIGURE 3-2 Principal job work activities question in the National Survey of College Graduates and Survey of Doctorate Recipients.
SOURCE: National Survey of College Graduates Questionnaire. Available: https://www.nsf.gov/statistics/srvygrads/surveys/srvygrads-newrespond2015.pdf [December 6, 2017].

**The next questions ask about your work activities in this position.
While in this position [have/did] you...**

(Select one response per row)

C41a: [Written/Write] technical reports or working papers other than those presented at conferences or submitted for publication?

C41b: [Attended/Attend] professional conferences when you were not presenting papers?

C41c: [Served/Serve] as a reviewer for a conference, journal, or an agency, institution, or organization?

C41d: [Supervised/Supervise] graduate students, postdocs, or research staff?

C41e: [Taught/Teach] courses?

C41f: [Developed/Develop] marketable products?

C41g: [Worked/Work] with researchers not a part of your department or lab?

C41h: [Served/Serve] as the principal or co-investigator on funded grants?

C41i: [Prepared/Prepare] grant or fellowship proposals in which you [are/were] listed as the principal or co-investigator?

C41j: [Prepared/Prepare] grant proposals in which you [are/were] **not** listed as the principal or co-investigator?

Administered to: All ECDs eligible for the iteration

1 = Yes
0 = No

FIGURE 3-3 Principal job work activities question in the Early Career Doctorates Survey (Web survey script).
SOURCE: Early Career Doctorates Survey Questionnaire. Available: https://www.nsf.gov/statistics/srvyecd/surveys/srvyecd_2015.pdf [December 6, 2017].

consider useful to have learned. The skills that are valued in the workforce are changing more rapidly than in the past, and there is a need to track these skills continuously (Golde and Dore, 2001; Rudd et al., 2008). Some of the skills needed in the science and engineering workforce (for example, management skills, managing a diverse workforce with more underrepresented groups, managing intellectual property rights) are not traditionally viewed as science and engineering skills. The need for this information is especially critical in the case of nonacademic careers, particularly in industry.

Workshop participants indicated that traditional academic institutions want to learn more about skills needed as input to their program planning (Aanerud et al., 2006; see also Chesley, 2015). Many are experimenting with integrating such approaches as targeted workshops, internships, training periods abroad, and collaborative projects, but often must speculate about what may be useful. Interdisciplinary programs also are on the rise, and there is increasing interest in understanding how these programs affect career trajectories (Martinez et al., 2015).

The Council of Graduate Schools is in the process of developing a survey of alumni to understand how they are using their skills in their careers and to provide feedback to graduate schools. The survey is designed to complement data collected by the SDR and SED, as well as data compiled by the Association of American Universities Data Exchange and individual institutions.

An overall question is whether the degree itself was useful and whether, looking back, the person would make the same decision to pursue a degree in the same field and at the same level (Aanerud et al., 2006).

In addition to interdisciplinary programs and degrees at traditional institutions, several government programs have been launched to further interdisciplinary training and research and facilitate collaboration across disciplines and internationally. These include the National Science Foundation's (NSF) Research Traineeship Program, the Partnerships for International Research and Education (formerly the Integrated Graduate Research Traineeship), and the National Institutes of Health's interdisciplinary programs. If the data from the SED and SDR could be linked to administrative records from these programs, this could greatly expand understanding of the impact of these programs (see below for detailed discussion of data linkages).

Community colleges and apprenticeships have received some attention among the less conventional sources of training. Many university extensions also are offering certificate trainings to help workers maintain or upgrade their skills or transition to a skill that is in high demand. In addition, major universities are increasingly making more courses, and in some cases master's degrees in science, technology, engineering, and mathematics (STEM) fields, available on the Internet.

Also seen over the past few years has been a rapid expansion in more informal sources of training. These include free online courses, such as those offered by Coursera, Udacity, and EdX; YouTube tutorials; Webinars offered by the journals *Science* and *Nature*; Websites that facilitate offline meetings, such as MeetUp.com; and even online communities, such as Stack Overflow and Stack Exchange. These types of resources can address the need to continuously update skills that characterizes some science and engineering fields, such as computer science, and also help expand participants' personal networks.

While the NCSES surveys ask about job-related training, the questions are focused on formal training (see Figure 3-4). Some forms of training, such as nondegree courses and especially informal training, including online tutorials and communities, may be missed. The surveys also ask about respondents' reasons for pursuing the training, and the items used

C1. During the past 12 months, did you take any work-related training, such as workshops or seminars?

Include conferences or professional meetings only if you attended a training session at the conference or meeting.

Do not include college coursework for which you were enrolled in a degree program.

1☐ Yes

2☐ No ⟶ ***Go to page 8, question C4***

C2. *(If Yes)* **For which of the following reasons did you take training during the past 12 months?**

Mark Yes or No for each item.

		Yes	No
1	To improve skills or knowledge in your current occupational field	1☐	2☐
2	To increase opportunities for promotion or advancement in your current occupational field	1☐	2☐
3	For licensure or certification in your current occupational field	1☐	2☐
4	To facilitate a change to a different occupational field	1☐	2☐
5	Required or expected by employer	1☐	2☐
6	For leisure or personal interest	1☐	2☐
7	Other – *Specify*	1☐	2☐

FIGURE 3-4 Training questions in the National Survey of College Graduates and Survey of Doctorate Recipients.

SOURCE: National Survey of College Graduates Questionnaire. Available: https://www.nsf.gov/statistics/srvygrads/surveys/srvygrads-newrespond2015.pdf [December 6, 2017].

for this purpose capture the reasons for formal training reasonably well (refer to Figure 3-4, question C2). Although researchers who study this topic would also be interested in additional details about the training, such as frequency, quantity, quality, satisfaction, and funding sources (e.g., paid for by the employer or employee), most of these details may be beyond what the NCSES surveys could accommodate.

RECOMMENDATION 3-3: To address the desire of researchers and policy makers for in-depth information on such topics as training and skills obtained and needed, the National Center for Science and Engineering Statistics should explore and evaluate adding to its surveys topic modules that would vary from round to round or be asked only of subsets of the samples.

EMPLOYMENT OUTCOMES, RESEARCH PRODUCTIVITY, AND ENTREPRENEURSHIP

Workshop participants called attention to the keen interest in measuring outcomes associated with various programs targeted at science and engineering students and the workforce. As a result, many disparate efforts are focused along these lines (Leggon and Pearson, 2008). Some academic institutions are following up with participants from their own programs in an attempt to assess the programs' impact, but these are small-scale efforts that could benefit from the availability of national data (Martinez et al., 2015). Also of great interest is being able to understand the outcomes associated with training, scholarship programs, fellowship programs, and internships among those who offer such programs, as well as in the federal government (e.g., outcomes associated with the Workforce Innovation and Opportunity Act).

Measuring outcomes is among the key goals for the NCSES science and engineering surveys, but one of the challenges is that outcomes, particularly those related to employment and careers (compared with those related to education), can be difficult to operationalize and measure (Rubio et al., 2011). The NSCG and SDR include widely used questions on labor force status and the relationship between education and occupation, as well as some questions on job and employer characteristics. The SDR also includes questions on academic position, faculty rank, and tenure status. The ECDS pilot study asked several questions about professional activities aimed at capturing productivity, including several follow-up questions about patents. These questions are shown in Figure 3-5.

The input obtained from workshop participants highlighted the need for these types of data on publications and patents. NCSES is already conducting research (described below) on possibilities for linking survey data to information about publications and patents. Questions related to factors that influence publications and patenting, such as collaborations and networks, would be useful to further understanding of these issues. In turn, the impact of scientific productivity on the ability to obtain research funding, promotions, or salary increases also is a topic of interest, particularly because of perceived gender differences in these areas (Blumenfield and Nerad, 2012; Nerad and Blumenfield, 2012).

In addition, researchers, practitioners, and policy makers want to better understand entrepreneurship, including involvement in launching a startup company. The NCSES surveys ask about whether the respondent is self-employed, capturing certain types of small businesses and successful entrepreneurs but not necessarily the founders of high-growth, innovative companies. The founders of startups would typically be more likely to consider themselves employees than self-employed, and thus would show up in the data as employees of a for-profit company. If it were possible to capture the concept of a startup, this information also would be useful for understanding the extent to which recent graduates are likely to be working at these types of companies versus other for-profit companies. This information would be useful not only for recent doctorate recipients but also for those without a graduate degree, and possibly even those with less than a completed bachelor's degree.

RECOMMENDATION 3-4: The National Center for Science and Engineering Statistics should continue to enhance its measurement of employment outcomes, research productivity, and entrepreneurship/innovation. Greater use of targeted survey modules and/or administrative and other nonsurvey data would enhance the feasibility and effectiveness of such measures.

WORK ENVIRONMENTS AND WORK–LIFE BALANCE

Motivation to choose a specific career or job and the decision whether to stay or leave are complex, and are heavily influenced by earnings opportunities, the nature of the work, and prospects for growth and advancement (Nerad, 2009). Such factors as illness, disability, military or other government service, and a voluntary move to part-time work can lead to changes in career trajectories as well. Also playing a role is the broader climate of the work environment, including the presence or absence of welcoming, talented, and supportive peers, supervisors, and mentors, as well as experiences of discrimination, harassment, or a hostile work environment, especially among members of historically underrepresented groups such as women, racial/ethnic minorities, and those with nontraditional gender identities.

The next questions ask about your work and research activities.

B1: Thinking about all of your paid, unpaid, and other research activities since receiving your [doctorate/doctorate-equivalent] degree in [field] from [institution], have you:

(Select one response per row)

B1a: Published papers in conference proceedings?
B1b: (Co)authored papers for presentation at regional, national, or international conferences? *(Do not include papers published in conference proceedings.)*
B1c: Had articles **accepted** for publication or already published in a peer-reviewed journal?
B1d: **Submitted** articles for publication in a peer-reviewed journal that were **not** accepted for publication or published?
B1e: Published books?
B1f: Published book chapters?
B1g: Been named as an inventor on U.S. patent applications?
B1h: Been named as an inventor on foreign patent applications?

1 = Yes
0 = No

Administered to: All ECDs Routing: If any one of B1a through B1h = 1 then go to B2; else go to B6

B2: How many...

B2a: Papers have you published in conference proceedings?
B2b: Papers have you (co)authored for presentation at regional, national, or international conferences?
(Do not include papers published in conference proceedings.)
B2c: Articles have you had **accepted** for publication or already published in a peer-reviewed journal?
B2d: Articles have you **submitted** for publication in a peer-reviewed journal that were **not** accepted for publication or published? *(If you have submitted one paper multiple times, count it only once.)*
B2e: Books have you published?
B2f: Book chapters have you published?
B2g: U.S. patent applications have you been named as an inventor on?
B2h: Foreign patent applications have you been named as an inventor on?

Figure continues on next page.

FIGURE 3-5 Questions about professional activities in the Early Career Doctorates Survey pilot study (Web survey script).
SOURCE: Early Career Doctorates Survey Questionnaire. Available: https://www.nsf.gov/statistics/srvyecd/surveys/srvyecd_2015.pdf [December 6, 2017].

Administered to: ECDs who have at least one work product since receiving their highest degree
Routing: If B2g > 1 or B2h > 1 then go to B3; else go to B6

B3: Since receiving your [doctorate/doctorate-equivalent] degree, how many patents have been granted to you as an inventor?

B3a: U.S. Patents Granted:
B3b: Foreign Patents Granted:

Administered to: ECDs who have been named as an inventor on a U.S. or foreign patent application since receiving their highest degree
Routing: If B3a > 1 then go to B4; else if B3b > 0 then go to B5; else go to B6

B4: Thinking of the U.S. patents that you have been granted since receiving your [doctorate/doctorate-equivalent] degree, for which of the following types of activity have they been granted?

(Select one response per row)

B4a: Utility patents (i.e., patents for inventions)
B4b: Plant patents (i.e., patents for fundamentally altered plant species)
B4c: Design patents (i.e., patents for appearance)

1 = Yes
0 = No

Administered to: ECDs who have been granted a U.S patent as an inventor since receiving their highest degree
----

B5: Since receiving your [doctorate/doctorate-equivalent] degree, have any of your patents resulted in commercialized products or processes or been licensed?

1 = Yes
0 = No

Administered to: ECDs who have been granted a U.S. or foreign patent as an inventor since receiving their highest degree

FIGURE 3-5 Continued

Considerations related to work–life balance greatly influence career choices and pathways as well. These considerations are sometimes related to childcare or eldercare responsibilities. In some cases, the decision to relocate is the result of a partner or spouse changing jobs (Goulden et al., 2009). These issues often affect women more than men, although there also are differences by race, in part because the sex ratio among those with advanced degrees varies by race (Malcom and Malcom, 2011; Ong et al., 2011; Ceci and Williams, 2010). These types of moves may decrease in the future as a result of the rise of telework options. In addition to information about the reasons for job changes, data on the educational background and occupation of spouses or partners would be useful to shed further light on these trends.

Researchers would like to know more about the characteristics of the workplace and work–life balance factors that can impact career paths. These include earnings, availability of mentoring, choice of projects, collegiality, collaborations, and telework options. A better understanding of the role of these considerations in career trajectories could advance research in this area. The new NCSES workforce survey, the ECDS, demonstrates responsiveness to emerging data needs in this area as well. The pilot study included a question about being away from work for 6 months or more, followed by a question about the reasons. The survey also asked about work stress level and about the ability to balance professional and personal responsibilities (see Figure 3-6).

Discrimination and harassment have long been recognized as barriers to full participation and performance to full capacity in science and engineering fields, not only for women but also for minority groups in general, yet no high-quality national data are available that could advance understanding of the extent, nature, and implications of these phenomena. This is an area NCSES has not yet explored, and although these topics are sensitive and difficult to measure, doing so is important to guide employer decisions and inform policy in this area.

RECOMMENDATION 3-5: The National Center for Science and Engineering Statistics should assess the performance of the new questions on working conditions and work–life balance in the Early Career Doctorates Survey and should consider adding these questions to the National Survey of College Graduates and the Survey of Doctorate Recipients as well.

RECOMMENDATION 3-6: The National Center for Science and Engineering Statistics should develop for the surveys core questions and a more in-depth module on harassment and discrimination.

D2: [Since you completed your [doctorate/doctorate-equivalent] degree in [field] from [institution]/Since you started your first full-time paid position], have you ever taken time off for six months or more?

1 = Yes
0 = No

Administered to: ECDs who received their highest degree in 2004 through 2013

Routing: If D2=1 go to D3; else if currently or previously employed go to D5; else go to E. Career Development and Plans

----

D3: Why did you take time off?
(Select one response per row)

D3a: For health reasons
D3b: To start a family or have more children
D3c: To care for a sick family member
D3d: To write papers from your dissertation for publication To travel
D3e: For additional education
D3f: You could not find employment
D3g: For some other reason (Please specify)

1 = Yes
0 = No

Administered to: ECDs who have taken off 6 months or more since receiving their highest degree or starting their first full-time paid position

----

D4: How many months did you take off?

Administered to: ECDs who have taken off 6 months or more since receiving their highest degree
Routing: If currently or previously employed go to D5; else go to E. Career Development and Plans

----

D5: Would you say that your [current/most recent] position [is/was]...

1 = Not stressful at all
2 = Somewhat stressful
3 = Very stressful
4 = Extremely stressful

Administered to: ECDs who have worked since earning their highest degree

----

Figure continues on next page.

FIGURE 3-6 Questions about principal job work activities in the Early Career Doctorates Survey pilot study (Web survey script).
SOURCE: Early Career Doctorates Survey Questionnaire. Available: https://www.nsf.gov/statistics/srvyecd/surveys/srvyecd_2015.pdf [December 6, 2017].

D6: Please indicate how much you agree or disagree with the following statements about balancing your [current] professional and personal responsibilities [when you were most recently employed].

(Select one response per row)

D6a: You [can/could] manage the demands of your position and at home.
D6b: Demands at home [limit/limited] your professional activities (e.g., going to conferences).
D6c: Your work schedule [allows/allowed] you to maintain the overall quality of life you want.
D6d: Your work schedule [provides/provided] the flexibility to take care of demands at home.
D6e: Your supervisor [understands/understood] when demands at home [interfere/interfered] with your professional responsibilities.
D6f: Demands at home [have] slowed down progress on your professional activities.

1 = Strongly disagree
2 = Disagree
3 = Agree
4 = Strongly Agree

Administered to: ECDs who have worked since earning their highest degree

FIGURE 3-6 Continued

POSSIBLE STRATEGIES FOR EXPANDING DATA ON THE SCIENCE AND ENGINEERING WORKFORCE

This section describes several possible strategies for expanding data on the science and engineering workforce and potentially addressing some of the key needs of data users. Because substantially increasing the length of the surveys would clearly not be possible without increasing respondent burden and jeopardizing response rates, several overall strategies could be considered that would likely have benefits in one or more of the areas discussed above.

Longitudinal Analysis

The designs of the NCSES workforce surveys include several rounds of follow-up with sample members. The NSCG currently interviews people for four cycles, and the SDR does so biennially until respondents reach the age of 76. However, data users generally cannot take advantage of the longitudinal aspects of the surveys because the data products are structured based on the model of a series of repeated cross-sectional surveys and do not include weights and documentation that would facilitate longitudinal analyses. This limitation has frustrated data users for many years. In addition, the recent SDR redesign led to the loss of about two-thirds of the existing panel cases in that survey (for a more detailed discussion, see Chapter 4).

As discussed earlier in this chapter and reflected in Recommendation 3-1, some analytic goals that NCSES considers priorities can be achieved with cross-sectional data, but changes in the science and engineering workforce are increasing the need to understand career pathways that are not as predictable and linear as they were in the past, as well as the impact of various factors, such as training and work environments, on career outcomes and job choices. There is also an increased need to better understand the inflows and outflows of scientists and engineers in the U.S. workforce, particularly international mobility.

Many of these emerging data needs could be addressed by taking better advantage of the longitudinal nature of the surveys—in other words, by treating the surveys as true panel designs that follow some of the sample members over the years. An increased emphasis on panel design also could enable a redesign of the questionnaires to focus on what has changed since the last contact with the sample member. This redesign could potentially free up space for a more nuanced understanding of the reasons for the changes, a gap in the current data.

Beyond integrating data from the different waves within each of the surveys, integration that enables longitudinal analysis across surveys would facilitate additional research. Measures related to interdisciplinary education that are included in the surveys are limited (field of degree and second field of degree in the NSCG, and field of interdisciplinary dissertation in the SED). The ability to conduct analyses on a dataset that longitudinally integrated responses from the SED and the different waves of the SDR would help in understanding how multidisciplinary education shapes future career pathways.

NCSES recently launched several activities that support longitudinal analyses. These activities include commissioning analytic work to construct properly weighted longitudinal data files for the past few rounds of the SRD and NSCG, and consulting with survey design experts and data users on design options for building up a substantial longitudinal component in the SDR going forward (see Chapter 4). Enabling longitudinal analyses would be the most far-reaching enhancement for expanding the usefulness of the NCSES workforce surveys, as reflected in Recommendation 3-1 above.

Survey Core Questions and Modules

One way to balance the need for fine-grained analysis of the core questions of the NCSES surveys with the need for occasional detail on some issues would be to redesign the surveys to accommodate topic modules. As discussed further in Chapter 5, this redesign would involve identifying a short list of core questionnaire items that would be administered to all sample members during every round of the survey, and grouping other questions into topical modules that would vary from round to round or be asked every round, but only of a subset of respondents. NCSES could develop survey modules that would serve as in-depth follow-ups to specific changes and perhaps be triggered by specific survey responses or information about a change from other sources of data, such as administrative records or other forms of nonsurvey data (discussed later in this chapter).

To some extent, the NSCG and SDR already rely on questionnaire structures that are based on core topics and topic modules administered periodically and focused on specific policy issues or on the gathering of in-depth information on a subpopulation. Over the years, both the NSCG and SDR have included modules on job satisfaction and attributes, federal support of work, immigration information, organization of work, international collaboration, and productivity measures (publications, patenting). In addition, the NSCG has included modules on professional certifications, community college enrollment, education debt (amount borrowed/owed), and academic positions. The SDR has included a module on postdoctoral experiences (see Chapter 2).

Recommendations 3-2 through 3-6 above identify topic areas in which there is a clear need for additional in-depth data that could potentially be addressed through topic modules: (1) factors that lead to career or job changes, including immigration to or emigration from the United States; (2) training and skills (including skill gaps); (3) employment outcomes and research productivity; (4) entrepreneurship and innovation; and (5) working conditions, including work–life balance and discrimination and harassment.

It is important to acknowledge that the desire of researchers and other data users for detailed information in many areas surpasses NCSES's ability to address these needs, even with greater reliance on topic modules. As discussed earlier, modules often increase respondent burden, particularly if many of the existing questions would be considered core questions that needed to be asked of everyone. To address this issue, it might be possible to experiment with administering a series of modules over a short period of time (some of which could consist of core questions and other questions that applied only to respondents with certain characteristics), and then evaluating how the response distributions and respondent perceived burden compared with the results from a longer version of the survey administered all at once (West et al., 2015). The use of modules for subsets of the sample also can greatly increase the complexity of a survey. Nonetheless, greater reliance on modules would be worth exploring, looking ahead to a system of longitudinal data that might integrate data across the surveys and from other sources. Increasing reliance on electronic modes of data collection, particularly Web administration, and a move away from paper-and-pencil questionnaires (see Chapter 5) also will likely make the use of customized modules more feasible.

NCSES could explore collaborations with other government agencies to support in-depth modules on topics of mutual interest and to identify efficient ways of collecting data on these topics. Agencies that might be particularly interested in detailed data on selected topics of mutual concern include the Bureau of Labor Statistics and the Census Bureau.

Other Questionnaire Modifications

In some areas, minor modifications to a questionnaire may be necessary or desirable to capture new characteristics and trends in the workforce. One question that warrants particular attention is the employer question on the NSCG and SDR, which asks for the employer's location (refer to Figure 3-1). Although the questionnaires collect address information for the respondent's residence separately, the physical location of the employer is important to understanding mobility, and the growing complexity of work arrangements may be making this question increasingly more challenging to answer.

The NSCG and SDR also ask respondents whether they were working for the same employer and in the same job during the reference week as during the reference week of the previous data collection. If the data were linked longitudinally, this information would be known, along with information about the location of the previous employer. The increased flexibility afforded by electronic data collection modes means there are fewer justifications for one-size-fits-all questionnaires, and in the future, information about a job change could drive subsequent questions, potentially reducing respondent burden and enabling NCSES to ask more detailed follow-up questions about the reason for the change. Currently, respondents who indicate that they changed either their employer or their type of job are asked to provide the reason. One of the answer options is "job location," but this response by itself does not enable an in-depth understanding of the mobility-related reasons. Reviewing the verbatim answers recorded under "some other reason" with the goal of identifying emerging trends could shed some light on whether this question could be revised or expanded.

Another question that could benefit from updating is that on training (refer to Figure 3-4). This question could be revised to ensure that it captures a broader range of training, particularly newer forms of informal training.

Integration of Data from Multiple Nonsurvey Data Sources

The NCSES surveys are essential for producing nationally representative data on key topics, many of which can be assessed meaningfully only through self-report because of their subjective or idiosyncratic nature. However, alternative data sources becoming increasingly available capture information of interest about scientists and engineers. As discussed earlier, some of these nonsurvey data sources are promising mechanisms for supplementing, or possibly even replacing, survey questions to reduce respondent burden.

NCSES already has some experience with the use of administrative data to enhance survey data collected by the SED. For the SED, institutional coordinators are asked to assist in obtaining from the institution's administrative records such missing critical items as year of birth, sex, race/ethnicity, citizenship, country of citizenship, bachelor's degree year and institution, and postdoctoral location. SED survey data also have been linked to administrative data from SSA to derive short- and long-term stay rates.

In addition, NCSES is exploring the use of alternative data sources to enhance information on such topics as employment outcomes, research productivity, and entrepreneurship. For the ECDS, NCSES is considering the use of data extracted from respondent-provided resumés and information from ORCID, a nonprofit organization that links education and research data for those who register. NCSES also is investigating the linkage of STAR METRICS, UMETRICS, and PROQUEST data to SED respondent data. Publication data from Web of Science and patent data from the U.S. Patent and Trademark Office are being linked to SDR respondent data as part of a feasibility study to evaluate data quality. NCSES is also experimenting with Web scraping to collect degree and work histories for SDR respondents. ScienCV could be another potential source of data to evaluate.

Some of the research on alternative data sources is being conducted in collaboration with the Census Bureau's Center for Administrative Records Research and Applications (CARRA). CARRA's research is focused primarily on data sources that could enhance the sampling frame for the NSCG, the survey for which the Census Bureau serves as NCSES's data collection contractor, but the research is likely to have implications for the other surveys as well. (Research on the use of alternative data sources to enhance the NSCG sampling frame is discussed in detail in Chapter 4.) The data sources being evaluated by CARRA include the Census Bureau's own Master Address File and Longitudinal Employer-Household Dynamics (LEHD) program; several other government agencies, such as the Internal Revenue Service (IRS), the Department of Education, and the Department of Homeland Security; career network Websites (such as LinkedIn, Monster.

com, and Chroniclevitae.com); online resumés; and the National Student Clearinghouse.

The potential use of data from the LEHD program, and unemployment insurance earnings data in general, is particularly promising because these data could serve multiple purposes. The Census Bureau's LEHD program combines state data with administrative data from federal agencies and from censuses and surveys to create a longitudinal database of linked employer–household data. Linking this information to IRS data, such as Business Register, Longitudinal Business Database, and Form W-2 data, could supplement the NSCG data with additional detail about sample members' employers and potentially provide a source of updates on employment that would reduce the need for frequent survey follow-ups. Unemployment insurance earnings data also could be linked to data from the Quarterly Census of Employment and Wages, which can identify new firms. Individuals taking new jobs at new firms could then be identified and receive targeted questions, for example, on entrepreneurship. One challenge associated with IRS data is that employer location and industry data are at the firm rather than the establishment level, which limits the ability to determine workplace location and type of work. Appendix D examines the possible use of unemployment insurance earnings data in further detail.

In terms of data available from other government agencies in general, one of the main challenges is related to developing procedures for and negotiating the terms of data sharing. Different surveys use different language to describe to participants how the data will be used and to obtain informed consent, which often limits how the data can be shared. Changes to the language must be carefully considered because of the possible impact on response rates. Another difficulty hindering collaboration between government agencies to advance work in this area is resource limitations, especially the fluctuation in resources available for this type of work within agencies. In the case of data from state agencies, the need to negotiate a data sharing agreement and adjust processes for each individual agency greatly increases the challenges. Nonetheless, some of the most valuable data for potential enhancement of the science and engineering surveys originate from other government agencies, and research on issues related to the use of nonsurvey data lends itself well to a variety of productive collaborations, such as the work being conducted with CARRA. These collaborations could be expanded to encompass methodological research (beyond the work conducted to evaluate data sources), building on the extensive methodological research experience of both the Census Bureau and NCSES. A variety of initiatives now under way government-wide could facilitate this type of collaboration going forward (see National Academies of Sciences, Engineering, and Medicine, 2017).

One limitation that needs to be addressed in assessing the potential use of these types of data is the lack of representativeness (in other words, data are likely to be available for some sample members and not others). In fact, researchers sometimes use NCSES survey data to benchmark data from other sources and to better understand the latter data's strengths and weaknesses. However, datasets do exist in which representation is very high for some types of scientist or engineer subgroups. On some topics, such as the types of projects on which a person is working, small-scale studies could be useful, focusing, for example, on the development of a richer dossier with substantial detail on the subset of sample members for which data are available. In some cases, data from external sources also could provide important information for research on topics related to the science and engineering workforce even if data from those sources could not be linked to the survey data.

Recent research in the area of record linkages has focused on ways of obtaining consent for linking survey data to data from other sources, as well as on the implications of consent bias, which is another source of concern in terms of the representativeness of the data from nonsurvey sources (Sakshaug et al., 2015, 2017). Privacy and confidentiality issues need to be considered and researched in general, in the case of all data from nonsurvey sources, including online databases and the Internet.

While due caution is necessary regarding the use of nonsurvey data, and despite the implementation challenges, options available for expanding the NCSES survey data with data from alternative sources are growing rapidly. Some of the identified gaps in the survey data are of the type that could be addressed through supplementation with data from other sources, and use of these types of data also could reduce respondent burden and attrition. The populations of interest for the NCSES science and engineering workforce surveys are particularly well suited to the use of online sources for supplemental data because more information about these sample members than about the general population is likely to be available online.

NCSES could be at the forefront of research on the use of the new data sources and assessment of their quality. Possible mechanisms for jumpstarting innovative research on combining data from multiple sources include establishing a research center of excellence; collaborating with academic centers; and using fellowships, dissertation awards, or grants to support research conducted by others.

The panel's recommendations below are consistent with the direction endorsed by recent reports of the National Academies of Sciences, Engineering, and Medicine (2017) and the Commission on Evidence-Based Policy Making (2017).

RECOMMENDATION 3-7: The National Center for Science and Engineering Statistics should explore mechanisms for accelerating research on the use of alternative data sources to expand and supplement the information obtained from the science and engineering workforce surveys.

RECOMMENDATION 3-8: The National Center for Science and Engineering Statistics (NCSES) should continue its collaboration with the Census Bureau toward expanding the use of external data sources to supplement its science and engineering surveys. A particular focus should be the use of unemployment insurance earnings data to supplement the survey data with additional, more complete, or more frequent data points regarding career trajectories. NCSES should also pursue collaborations with other government agencies to explore options for combining data from different sources.

INTERNATIONAL EFFORTS AND COLLABORATIONS TO EXPAND SCIENCE AND ENGINEERING WORKFORCE DATA

With an increased interest in international comparisons of science and engineering workforce data on particular topics such as international mobility, it is important for NCSES to continue its collaborations with international partners to coordinate data collection strategies and definitions. OECD, in collaboration with the United Nations Educational, Scientific and Cultural Organization Institute for Statistics, and Eurostat recently launched the Careers of Doctorate Holders project, which aims to develop internationally comparable indicators on the careers and mobility of doctorate recipients. The project was developed with support from the European Union and NSF. These types of collaborations are important because international comparability greatly enhances the usefulness of the data.

Interest in alternative data sources is increasing in other countries as well. For example, OECD is exploring the use of databases of inventors and authors of scientific publications. One project has examined the international mobility of scientists based on a mobility measure derived from data on affiliations obtained from a global index of scholarly publications (Appelt et al., 2015). Another OECD initiative, the International Survey of Scientific Authors, follows up on a stratified random sample of corresponding authors from the same publications database (Boselli and Galindo-Rueda, 2016; see also European Science Foundation, 2013, 2015).

Likewise, Statistics Canada has been experimenting with the use of administrative data from two government sources: the Postsecondary Student Information System (PSIS) and the Registered Apprenticeship

Information System (RAIS). PSIS is a registry of enrollments in postsecondary institutions containing several variables, including field of study, while RAIS contains similar information for registered apprentices. Over the past few years, PSIS and RAIS have become the backbone of an integrated system called the Education Longitudinal Linkage Platform (ELLP). The ELLP complements the PSIS and RAIS data with data from other sources, in particular, income tax data. More important, this linkage is maintained longitudinally, enabling researchers to track the paths of individuals through the education system and then the workforce. There are plans to add other data sources to the ELLP, such as student loan files and survey data.

A recently published pilot study using the ELLP for graduates from universities in Canada's Maritime provinces compares the economic outcomes of graduates who earned their degree between 2006 and 2008 with those of later cohorts (Galarneau et al., 2017). This project provides a good example of the type of longitudinal analysis that can be conducted by combining data from purely administrative sources.

RECOMMENDATION 3-9: The National Center for Science and Engineering Statistics should continue to consult and collaborate with international partners on ways to improve and enhance data on the science and engineering workforce.

Although the availability of data sources varies by country, particularly in the case of administrative data, international collaboration and consultation can generate valuable ideas with potential to be adapted for implementation in the United States. Collaboration with the goal of researching the use of alternative data sources that may be internationally applicable, such as data extracted from online databases, could be especially useful.

MONITORING OF FUTURE DATA NEEDS

NCSES's science and engineering workforce surveys have remained current over the decades because of the agency's due diligence in keeping up to date on emerging workforce issues. However, the pace of change is increasing, and the emerging issues are increasingly more complex and substantially more difficult to measure. To meet this challenge, NCSES needs to incorporate into its planning efforts mechanisms that will allow it to remain well informed about stakeholders' data needs and emerging topics of importance.

One option would be to involve an advisory board focused on providing guidance on the prioritizing of topic modules. The NSF Directorate for Social, Behavioral, and Economic Sciences has an advisory committee, but it does not always include members with expertise specific to NCSES's mission. Another option would be to issue calls for proposals for new topics, as done for the General Social Survey. This mechanism also could facilitate broader use of the surveys in general because more researchers would become familiar with the core questions included in the surveys on a regular basis.

Another option would be to communicate more extensively with user groups. NCSES recently issued a task order to enhance outreach efforts for its Human Resources Statistics (HRS) Program by strengthening relationships with current customers and developing the groundwork for interaction with potential customers. The goals are to

- identify and document the current and potential HRS customers;
- improve HRS's understanding of the needs of its customers and how they interact with the HRS data products and other data sources;
- gather feedback on how the HRS survey content aligns with customers' data needs;
- strengthen the NCSES brand by increasing awareness of the HRS data and data products; and
- develop a framework for continued interactions with the HRS customer base.

The task order represents a promising effort to broaden understanding of stakeholder needs, and the focus on developing a framework for continued interactions with stakeholders is particularly important. A possible avenue for obtaining regular input would be to form a data user group that could serve as a sounding board on issues related to survey content. Periodic data user conferences or workshops would also be invaluable. Collaborations with academic centers, fellowships, dissertation awards, and grants would be other good means of keeping current on emerging topics of interest concerning the science and engineering workforce (for additional discussion, see Chapter 6).

RECOMMENDATION 3-10: The National Center for Science and Engineering Statistics should implement mechanisms for continuously monitoring input on emerging data needs from stakeholders in the academic, research, policy-making, and practitioner communities.

4

Sample Design and Estimation Procedures

As discussed in Chapter 2, the National Center for Science and Engineering Statistics (NCSES) conducts two annual census surveys to collect data on the education of scientists and engineers: the Survey of Graduate Students and Postdoctorates in Science and Engineering (GSS) and the Survey of Earned Doctorates (SED). The agency also conducts three sample-based workforce surveys: the National Survey of College Graduates (NSCG), the Survey of Doctorate Recipients (SDR), and the Early Career Doctorates Survey (ECDS). This chapter focuses on potential enhancements to the sample design and estimation procedures used for the latter three surveys.

NCSES'S WORKFORCE SURVEYS

NSCG

The NSCG is a biennial survey whose sample includes individuals who have a bachelor's degree or higher, are under the age of 76, and reside in the United States in a noninstitutional setting. Since 2010, the NSCG has used a rotating panel design with stratified sampling that combines samples from the most recent round of the American Community Survey (ACS) and the previous three rounds of the NSCG. As discussed in Chapter 2, the use of the ACS as a sampling frame for the NSCG was a recommendation made by a previous panel of the National Academies of Sciences, Engineering, and Medicine when the ACS replaced the decennial census long form (National Research Council, 2008). Prior to the transition to the ACS, NCSES used the long form for information on college graduates. The ACS collects data continuously (annual estimates are produced based on a series of monthly samples), which represents an advantage over the long-form data that were available only once a decade. Figure 4-1 shows the NSCG sample design, which includes four rounds of interviews and a typical expected sample size of about 130,000.

The variables used in the stratification process for the NSCG sample include highest degree type; occupation/degree type (a composite variable that captures both occupation field and bachelor's degree field); a composite demographic variable that combines information about race, ethnicity, disability status, and citizenship; and the likelihood that the person holds a degree obtained in the United States. Recent NSCG surveys have included an oversample of young graduates with a bachelor's or master's degree, which serves to replace data previously collected through the National Survey of Recent College Graduates (NSRCG).

The NSCG sample allocation is based on reliability targets for the survey's key analytic domains, which are determined primarily according to the needs of NCSES's two congressionally mandated reports, *Science and Engineering Indicators* and *Women, Minorities and Persons with Disabilities in Science and Engineering*. These primary data needs include characteristics defined by degree level, occupation field, degree field, demographic group, and being a young graduate.

SDR

The SDR is a biennial survey whose sample includes individuals who have received a research doctoral degree in a science, engineering, or selected health field from an academic institution in the United States are under the age of 76 and reside in a noninstitutional setting. Sample members are contacted every 2 years, and unlike the NSCG, the SDR since 2010 has followed up with sample members living both in the United States and abroad.

The SDR uses as a sampling frame respondents to the SED census of completed doctorates (which consists of approximately 55,000 new graduates annually, including about 38,000 graduates in science, engineering, and selected health fields). In 2015, the survey underwent a major redesign that involved a large increase in sample size, from typically around 40,000 cases prior to 2015 to around 120,000 cases. The goal of the redesign was to enable NCSES to produce estimates of employment characteristics at the fine field-of-degree level that is reported in the SED (215 fine fields of degree, compared with the 26 broad fields reported in the past), while at the same time maintaining the aggregate degree-level estimates for domains defined by demographic characteristics used in the main NCSES publications.

The 2015 sample redesign has engendered some controversy among some of the data users. From the panel's point of view, the redesign has

NSCG Year	Cohort/Panel: Returning Sample	Cohort/Panel: New Sample	Total
2010	2008 NSCG & RCG *(n=35,000)*	2009 ACS *(n=65,000)*	2010 Total: 100,000
2013	2010 RCG *(n=13,000)*; 2009 ACS *(n=47,000)*	2011 ACS *(n=83,000)**	2013 Total: 143,000
2015	2010 RCG *(n=12,000)*; 2009 ACS *(n=23,000)*; 2011 ACS *(n=60,000)*	2013 ACS *(n=40,000)**	2015 Total: 135,000
2017	2009 ACS *(n=23,000)*; 2011 ACS *(n=28,000)*; 2013 ACS *(n=25,000)*	2015 ACS *(n=48,000)**	2017 Total: 124,000
2019	2011 ACS *(n=28,000)*; 2013 ACS *(n=25,000)*; 2015 ACS *(n=30,000)*	2017 ACS *(n=48,000)**	2019 Total: 131,000

FIGURE 4-1 Sample design of the National Survey of College Graduates.
NOTES: *Indicates the inclusion of an oversample of young graduates. RCG refers to the discontinued National Survey of Recent College Graduates.
SOURCE: Finamore (2016).

both positive and negative aspects. Positive aspects are described in this section, negative aspects in the following section on Longitudinal Design. Regardless of the pluses and minuses, the panel sees good opportunities for further increasing the usefulness of the SDR going forward.

The sample size increase and the redesign were implemented at the request of the U.S. Office of Management and Budget and based on consultation with an expert panel to address a need for understanding the characteristics of the science and engineering workforce by specific fields of study. Many of the federal government's investments in research and advanced training are focused on specific fields of study, and the level of detail previously available by broad categories of fields of study was not always adequate to inform policy decisions.

The redesign enhanced NCSES's ability to produce reliable estimates of counts and employment characteristics of doctorate recipients by such subfields as "bioinformatics" and "biotechnology" in addition to the broader field of "other biological sciences"; to track changes by subfield; and to make this information available by such demographic characteristics as gender, ethnicity, and race. The redesign will enable analyses that will help explain variations within subfields that can inform policy decisions aimed at broadening participation by underrepresented groups, for example, by answering such questions as the following: What are the occupation category differences for different groups in particular subfields? Is there "early attrition" from the research track by some demographic groups? and Are women in particular academic disciplines being promoted/tenured at the same rate as men? The data also will shed light on the sometimes very different employment patterns of graduates in different types of subfields (for example, subfields focused on basic versus applied research) within a broader field. In addition, it will be possible to conduct research on foreign-born recipients of doctoral degrees in specific fields (for example, among foreign-born U.S.-trained doctorate holders in particular fields, who stays in the United States and who returns home, and the employment characteristics of foreign-born doctorate holders when working abroad or when working in the United States). NCSES expects the 2015 SDR public-use data file to expand the data available by field of degree from the 20 broad fields in the 2013 file to approximately 90 fields of degree. Although the sample sizes are still not large enough to enable NCSES to release data for all fine fields of degree in the public-use data file, the restricted-use microdata file that is made available to researchers who obtain a license will include all 215 fine fields of degree.

Prior to 2015, the survey used a fixed-sample panel design, which meant that new graduates were added to the sample in each survey cycle, and sample members were dropped only when they no longer met eligibility criteria or if doing so was necessary to maintain the overall sample size. With the 2015 sample, NCSES attempted to optimize weighting and variance estimation in 215 fine field-of-degree strata for demographic characteristics. The downside of this design focused on fine fields of degree is that it entails highly unequal sampling rates across strata, creating some inefficiencies (discussed later in this chapter).

For the 2015 sample, NCSES considered a complete recasting of the sample, based on the recommendation of a panel of experts asked to assist with the redesign. In the end, the agency decided to drop approximately two-thirds of the existing cases, but carry forward 16,075 cases from the eligible 2013 sample to enable analysis using these older cases. Still, the majority of the redesigned sample is new to the SDR. Table 4-1 shows the 2015 sample by cohort. In the 2017 SDR, NCSES carried forward the entire 2015 sample to facilitate an improved longitudinal design for this survey in subsequent cycles of data collection. This approach is consonant with the emerging data needs and recommendations presented in Chapter 3, whereas the 2015 design was not.

ECDS

As discussed in Chapter 2, the ECDS is a new survey, for which a pilot data collection was conducted in 2014–2015. The pilot sample included individuals who had earned their first doctorate or doctorate-equivalent degree within the past 10 years and who were working at an academic institution included in the GSS, in a federally funded research and development center (FFRDC), or in the National Institutes of Health Intramural Research Program (NIH IRP). The pilot survey had a two-stage stratified sample design. Primary stratification was based on the above three sectors—academic institution, FFRDC, and NIH IRP. Academic institutions were further stratified by type: highly active research institutions, institutions with medical schools, doctorate-granting institutions, and master's-granting

TABLE 4-1 Sample of the 2015 Survey of Doctorate Recipients (SDR), by Cohort

Sample Cohort	2015 Count
Panel continuing from 2013 SDR	16,075
Expansion sample: new to the SDR in 2015	87,252
Expansion sample: previously deselected and resurrected for 2015	6,336
New graduates sample (academic years 2012 and 2013)	10,337
Total	120,000

SOURCE: Finamore (2016).

institutions. Academic institutions were selected with probability proportional to size, while FFRDCs and the NIH IRP were sampled with certainty. After the institutional data collection had been completed, individuals were selected from the institutions that responded. Stratification was based on the likelihood that the person was an early career doctorate recipient and postdoctoral status. Approximately 44 sample members were selected from each of the academic institutions and FFRDCs, and 250 individuals were sampled from across the NIH IRP.

Possible Enhancements to Sample Design and Estimation Procedures

As discussed above, the sample designs for the NSCG and SDR are based on achieving prespecified precision requirements for a large number of respondent categories. For example, the SDR has specific sample sizes allocated to 216 strata, defined by fine field of doctoral degree. The sample sizes for these strata range from 3,661 (with a sampling fraction of 0.09) to 42 (with a sampling fraction of 1.00). The overall sampling fraction is about 0.11. The NSCG sample design is similarly motivated.

The precision requirements are carefully specified in the design documents for the surveys, and it is obvious that great attention has been paid to ensuring that the respective designs meet the goals for adequate precision of results for a large group of reporting categories. Accordingly, it would appear that two primary considerations apply to potential enhancements to the sampling design and estimation procedures for these surveys:

1. What enhancements would permit the current design goals to be met more efficiently?
2. Are there other design goals that should be considered, either in addition to the current goals or requiring some kind of compromise with the current goals?

The remainder of this chapter reviews possible enhancements to both sample design and estimation that address these considerations.

SAMPLE DESIGN

Longitudinal Design

In principle, both the NSCG and SDR are longitudinal panel surveys, but historically NCSES has focused on the surveys' primary goal of producing cross-sectional estimates and maintained the longitudinal sample design simply to take advantage of cost efficiencies associated with interviewing the same sample members over multiple cycles. NSCG sample members are interviewed four times, and most SDR cases remain in the sample until the age of 76.

Although some researchers have created longitudinal datasets based on these surveys in the past, the data files released by NCSES were not configured to facilitate this type of use and did not include weights appropriate for longitudinal analyses. The cuts made to the SDR sample to maintain the overall sample size (discussed earlier) also introduced challenges over the years.

Despite the fact that some cases from previous years were maintained in the new sample even after the 2015 SDR redesign, the changes involved the loss of about two-thirds of the existing cases. This represents a substantial setback for potential longitudinal uses of the data, but also an opportunity to develop a design that better supports these types of analyses because the design can be developed with longitudinal analysis goals in mind.

Following the redesign, NCSES launched several activities focused on enhancing the usefulness of the data for longitudinal research going forward. First, the agency commissioned technical and analytic support to construct properly weighted longitudinal files for the 2008, 2010, and 2013 SDR, as well as the 2010, 2013, and 2015 NSCG. This work is ongoing, and it is expected to serve as a template for longitudinal files based on data from subsequent survey cycles as well. NCSES also is working with a panel of experts in sample design to inform longitudinal design options for the 2019 SDR cycle and has been proactive in informing the data user community about the options that are being considered and the progress of the work.[1] To obtain input before finalizing the plans, the agency has scheduled a meeting with stakeholders in August 2017 and is planning to hold an additional meeting in early 2018.

As discussed in Chapters 1 and 3, the present panel believes enabling longitudinal analyses is the most important enhancement for expanding the usefulness of the NCSES workforce surveys, particularly to meet increasingly urgent data needs to better understand career pathways and mobility (see Recommendation 3-1 in Chapter 3). The work on the SDR sample design is promising because the agency has made demonstrated progress in the direction of building up a longitudinal sample in the SDR: in addition to the activities described above, it carried forward the 2015 sample to the 2017 SDR. The panel assumes that the design changes going forward will also address the previous inefficient practice of ad hoc cuts to maintain sample size.

---

[1]Summaries of the expert panel meetings are available at https://www.nsf.gov/statistics/srvydoctoratework/longitudinal-design.cfm [October 27, 2017].

RECOMMENDATION 4-1: The National Center for Science and Engineering Statistics should maintain and accelerate its efforts to develop sample designs that facilitate the increased generation and analysis of longitudinal data in the National Survey of College Graduates and the Survey of Doctorate Recipients, including the generation of appropriate weights and documentation for longitudinal analysis.

If integration with data from administrative records, such as unemployment insurance earnings data, becomes more feasible in the future, this development could pave the way for a longitudinal system of surveys in which some of the updates would originate from administrative databases or from custom follow-up surveys triggered by the updates to administrative data (see Appendix D for a more detailed discussion).

International Mobility of Doctorate Recipients

The goal of the SDR has been to provide data on the employment characteristics and career pathways of individuals who obtained a doctorate in a science, engineering, or selected health field from an academic institution in the United States, and sample members have been interviewed even if they moved abroad after obtaining their degree. With the 2015 redesign, the sample became fully representative of these graduates, regardless of where they reside at the time of the survey. This change allows the SDR to capture important populations that represent substantial subsets of U.S. doctorate recipients. The importance of these groups of doctorate recipients is demonstrated by data from the SED indicating that in recent years, approximately one-third of doctorates awarded by U.S. institutions in the fields of interest have been awarded to individuals with a temporary visa, and about one in four of these degree recipients with a temporary visa were planning on leaving the United States after graduation (National Center for Science and Engineering Statistics, 2017e). The expanded sample enables analyses of the employment characteristics of foreign-born recipients of degrees from U.S. institutions by fine field and by residence in the United States or abroad. Figure 4-2 shows for the 2013 and 2015 SDR the number of sample members who reside abroad, by degree year.

Scientists and Engineers with Foreign-Earned Doctorates

One group of doctorate recipients not well represented in the current system of surveys is those who obtained their degree abroad but are now living in the United States. The NSCG samples individuals currently residing in the United States, and as discussed, the variables used in the stratification process include a composite demographic variable combining information on

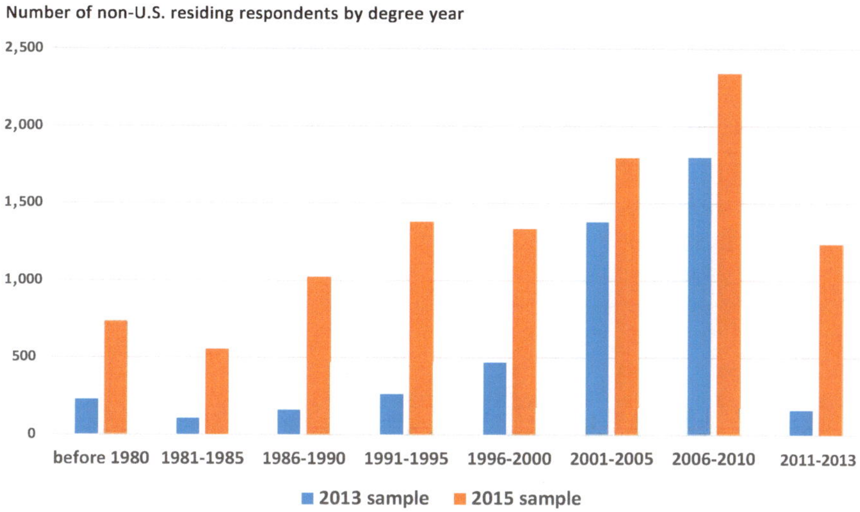

FIGURE 4-2 Number of Survey of Doctorate Recipients sample members residing abroad by degree year, 2013 and 2015.
NOTE: The sample size was increased from approximately 40,000 in 2013 to approximately 120,000 in 2015. The 2013 SDR included only eligible graduates from the 2011 academic year and did not fully represent those living abroad who earned their degrees prior to academic year 2001.
SOURCE: National Center for Science and Engineering Statistics (2017f).

race, ethnicity, disability status, citizenship, and the likelihood that the person holds a degree obtained in the United States. However, the information available from the ACS is not ideally suited to obtaining a large enough sample of this population for the NSCG to make the survey truly useful for tracking the international flow of doctorate recipients with degrees obtained abroad. In the 2015 NSCG, for example, the number of holders of foreign-earned doctorates who originated from the 2013 ACS sampling frame was 243. In other words, after 4 years, the rotating panel design that combines samples from the most recent round of the ACS and the previous three rounds of the NSCG is expected to include around 1,000 foreign-earned doctorates.

The sample of doctorate recipients with degrees obtained abroad could be increased in the NSCG in three ways. First, it might be possible to improve the stratification so that oversampling would be more efficient in identifying those with foreign-earned doctorates. Second, it might be possible to retain members of this group in the survey sample for a longer period of time. The third option is expanded use of data from the ACS for NSCG sample design and estimation. These options are discussed below. With all of these options, in addition to generating more adequate sample sizes of doctorate recipients with degrees from abroad, the NSCG would follow sample members who moved abroad, regardless of degree level.

Improved Stratification to Increase the Sample of Foreign-Earned Doctorates

The ACS does not collect information about when and where respondents earned their degrees, so for sampling purposes, NCSES assumes that people born in the United Sates have a "high likelihood of a U.S.-earned degree." For foreign-born individuals, NCSES uses information on age, year when the person came to live in the United States, and highest degree attained to develop a proxy variable for the likelihood of a U.S.-earned degree (the ACS does not ask for year of degree). In addition to not measuring whether a degree was earned in the United States, the ACS questions have limitations as proxy variables. In the case of individuals who came to live in the United States on more than one occasion, for example, the ACS asks only about the most recent year in which the person moved to the United States. For NSCG sampling purposes, it would be more useful to know the year when a person first moved to the United States (or ideally whether the person's degree was earned outside of the United States).

Using the proxy information available from the ACS, NCSES categorizes foreign-born individuals with a doctorate as their highest degree as having a "high likelihood of a U.S.-earned degree" if they came to live in the United States when they were aged 24 or younger. If they came to live in the United States at age 25 or older, they are categorized as having a "low likelihood of a U.S.-earned degree." The majority of sample members classified as having a "low likelihood of a U.S.-earned degree" on the basis of their responses to the ACS also report a foreign-earned degree in the NSCG, but NCSES finds that many individuals have a foreign-earned bachelor's or master's degree and a U.S.-earned doctorate. In the 2013 and 2015 NSCG samples, the stratum of doctorate recipients with "a low likelihood of a U.S.-earned degree" included around 50 percent actual foreign-earned doctorates, around 40 percent actual foreign-earned bachelor's or master's degree holders who had no foreign-earned doctorate, and approximately 10 percent without a foreign-earned degree.

The variables used to identify the strata (age, highest degree, and year when the person moved to the United States most recently) are, in all likelihood, the strongest predictors of foreign-earned doctorates. However, it might be possible to improve the sensitivity of this classification by using other data available from the ACS, such as birthplace, citizenship, or veteran status (see the next section). Several modeling techniques would be appropriate for this classification task, such as random forest models, gradient-boosted decision trees, logistic regression, or classification and regression trees (CART) (Hastie et al., 2009).

A suitable modeling approach would use respondent data from the NSCG to define the outcome (foreign-earned doctorate), with the predictors coming from the available ACS data. A logistic regression modeling approach, for example, would predict the NSCG-reported degree (highest foreign-earned degree is a doctorate, 1 = yes, 0 = no) as the outcome. The data from the ACS could be used as predictors. The current rule could be emulated, for example, by including a predictor that coded cases as meeting the current rule for inclusion in the "low likelihood of a U.S.-earned degree" category (1 = meets rule, 0 = does not meet rule). In such a model, the estimated probability would likely be approximately 0.50 for cases in the group that met the rule. However, a logistic regression model has more flexibility to include more predictors. A CART model might also be very useful for this classification problem.

Given that the most likely predictors of a foreign-earned doctorate available from the ACS are used in the current classification scheme, the gains from using additional variables are likely to be relatively small. If the stratification can be improved, however, this would make it more efficient to increase the sample size for this population in the NSCG.

RECOMMENDATION 4-2: The National Center for Science and Engineering Statistics should evaluate whether modeling techniques could be used to improve the sensitivity of the classification used to stratify the National Survey of College Graduates (NSCG) sample drawn from the American Community Survey, thereby making it more efficient to increase in the NSCG the sample of doctorate holders who obtained their degree from abroad.

Longer Retention of Identified Foreign-Earned Doctorates

Given interest in the movement in and out of the U.S. workforce of doctorate recipients who obtained their degree abroad, it may be useful to track these cases for longer than the four waves of the current NSCG design. If the longitudinal analysis capabilities of the survey can be enhanced, the NSCG panel design would be particularly well suited to studying the within-person movements of this population in and out of the U.S. workforce. Additional follow-ups also would have the advantage of increasing the sample sizes for this group as more cases were accumulated over time.

The decision to limit the number of follow-ups in the NSCG to four was based on the observed pattern of decline in response rates in subsequent waves. However, the SDR manages to follow sample members for longer periods. Furthermore, if it is particularly important to understand the mobility of the NSCG population, it may be worth considering alternative design options or customized survey procedures for this group (although these types of approaches would be applicable to any group with specific characteristics and a small sample size or the entire NSCG sample if the number of follow-ups were increased). Such alternatives could include shorter survey instruments, more frequent administration of shorter instruments, sending between-wave updates to panel members, higher incentives, tailored contact materials, or more intensive locating efforts. It might be possible to create a very short survey focused on key questions about this group that could be administered annually or biannually over a longer period of time. Reliance on administrative records for some of the data might also be an option.

In determining the optimum number of waves, it would be necessary to balance the goals of maximizing the benefits of the longitudinal design and maintaining reasonable response rates. The cost implications of the additional follow-up would have to be assessed as well.

RECOMMENDATION 4-3: To enhance understanding of the movements in and out of the U.S. workforce of those who obtained their doctorate degree abroad, the National Center for Science and Engineering Statistics should consider increasing the number of follow-up waves for this subgroup in the National Survey of College Graduates.

Expanded Use of Data from the ACS for NSCG Sample Design and Estimation

As discussed, the ACS variables (or variables derived from the ACS) currently used in the stratification of the NSCG sample include highest degree type; occupation/degree type (a composite variable that captures both occupation field and bachelor's degree field); and a composite demographic variable combining information about race, ethnicity, disability status, citizenship, and the likelihood that the person holds a degree obtained in the United States. The ACS collects additional data that do not appear to be used in the NSCG, but could potentially be useful at either the design or estimation stage. These data include geographic location, gender, age (more generally than just defining young graduates), birthplace, veteran status, employment status, employer type, industry, and income.

These additional variables could be used in one or more of three ways. First, additional information from the ACS could be used at the design stage, either for explicit stratification, for implicit stratification (sorting), or to determine selection probabilities. For example, it would appear that the process of identifying those who obtained a doctorate outside the United States might be enhanced by including birthplace and veteran status in the model. Information obtained from the ACS also could be used to enhance the NSCG data file, particularly in the case of variables not subject to change over time. This type of potential use might have confidentiality implications, but researching its feasibility would be worthwhile. Finally, ACS data could be used in estimation, for calibration or generalized regression. Although the benefits of using the same data for design and estimation may be limited, doing so could still be effective when nonresponse is substantial, as discussed in more detail below.

RECOMMENDATION 4-4: The National Center for Science and Engineering Statistics should evaluate potential additional uses in the National Survey of College Graduates (NSCG) for information obtained from the American Community Survey (ACS). These additional uses could include using additional variables in the design stage, using ACS data to enhance the NSCG data file, and using ACS data in estimation.

Possible Alternative Sampling Frames for the NSCG and SDR

As discussed in Chapter 3, NCSES is researching the use of alternative data sources, such as administrative records, to expand and improve the range, accuracy, and efficiency of the data collected through its science and engineering workforce surveys. Alternative data sources also could be considered for use in the sampling frames for the surveys, either as supplemental frames, combined with samples from the current frames (the SED for the SDR or the ACS for the NSCG), as part of multiple-frame methods, or if necessary as replacement frames. The SED and ACS, respectively, represent the most suitable sources of data for the SDR and NSCG sampling frames, and the potential loss of the SED or ACS as data sources for the respective sampling frames would likely lead to a substantial loss of precision for the affected survey. However, continued research to evaluate alternative data sources appears prudent in case either of the current sampling frames should become degraded or unavailable.

NCSES is working with the Census Bureau's Center for Administrative Records Research and Applications (CARRA) to evaluate these types of alternative frame options. CARRA's research is focused on the NSCG because the Census Bureau is the current data collection contractor for that survey. It is therefore more feasible to link the NSCG, relative to the other NCSES surveys, to other data that are already available or can be obtained by the Census Bureau. However, some of the alternative data sources being considered by CARRA could also be useful for the other NCSES workforce surveys. The main alternative data sources being actively researched by CARRA on behalf of NCSES are described below, along with the panel's perspectives on their potential. As stated in the panel's charge, NCSES is particularly interested in the use of multiple data sources, but is cognizant that the availability of these data sources can change over time.

Federal and Other National-Level Data Sources

Form 1098T data. One type of Internal Revenue Service (IRS) data that could be used for the sampling frame of NCSES's science and engineering surveys is data from form 1098T. This form includes tuition information and is filed annually by educational institutions and insurers that provide reimbursements or refunds of tuition and related expenses. The Census Bureau has an agreement with the IRS to obtain data on payments for qualified tuition and related expenses and scholarships or grants, along with information on whether the student was attending the institution at least half-time or was a graduate student.

The form 1098T data could potentially provide a frame of individuals who already have a bachelor's degree and are currently enrolled as graduate students. Useful information on the form is limited, but includes the institution where the student is enrolled and the student's place of residence. These data might be useful if there were a particular desire to oversample students currently enrolled in graduate degree programs.

The ACS asks about whether the respondent is currently enrolled in a graduate degree program, subdividing responses into master's, professional, and doctorate programs. Also available from the ACS is information on undergraduate field of degree, improving the ability to identify graduate students in fields of interest. Thus, the use of form 1098T data appears worth considering only if either ACS data became unavailable or the ACS sample size were determined to be insufficient for some future purpose.

W-2 tax form data combined with Business Register (BR) and Longitudinal Business Database (LBD) data. Another way in which IRS data could potentially be used for an NSCG frame would be to combine W-2 tax form data with data from the BR and LBD, which are databases of business establishments and companies maintained by the Census Bureau. Combining these data from these sources could result in a sampling frame containing extensive information about individuals' income and some characteristics of their employers, particularly if the W-2 could be obtained and included. However, addressing the confidentiality issues associated with the use of these data would likely be challenging, and information on educational background and current occupation would be imprecise.

Longitudinal Employer-Household Dynamics Program (LEHD) data. As discussed in Chapter 3, linkages to the Census Bureau's LEHD data could supplement the NSCG data with additional detail on sample members' employers and could potentially serve as a source of updates on employment that would reduce the need for frequent survey follow-ups. However, there are coverage limitations associated with the LEHD data (they do not include the self-employed), and the database contains insufficient information to be suitable for use in developing a sampling frame (information on educational background or occupation is lacking).

National Student Clearinghouse (NSC) data. The NSC is a nonprofit organization that provides educational reporting, data exchange, research, and degree and enrollment verification services. More than 3,600 colleges and universities, representing 98 percent of all students enrolled in educational institutions in the United States, participate in the clearinghouse, and the degree verification service encompasses about 90 percent of 4-year degrees obtained in the United States. In CARRA's assessment, the NSC database appears to show the most promise as a potential source of data for a sampling frame replacement, although obtaining access to NSC data needs to be viewed as a long-term project, primarily because of confidentiality restrictions as the Family Educational Rights and Privacy Act (FERPA) requires students' permission for uses of the data beyond evaluations of specific programs. In addition, rates of missing and underreported data in the NSC are higher for certain demographic groups.

U.S. Department of Education data. The National Student Loan Data System (NSLDS) is a database of student aid recipients maintained by the U.S. Department of Education. Again because of FERPA restrictions, there are limitations on how these data can be used. The Department of Education cannot provide its data files to the Census Bureau for use in constructing a sampling frame, but may be willing to provide limited information on specific individuals who are already in the NSCG sample, thereby facilitating follow-up between data collection waves.

Department of Homeland Security (DHS) data. DHS approves visa applications for entry into the United States, including student visas (F-1) and exchange visitor visas (J-1). The agency maintains information about foreign nationals, including education status and location and description of work in the case of J-1 visa applicants, that could be a source for a supplemental sampling frame on foreign nationals. However, DHS has not indicated willingness to share this type of data with the Census Bureau.

Other alternative data sources. As discussed in Chapter 3, NCSES, with the help of CARRA, is evaluating several other alternative sources of data, such as online resumés and information available from career networking sites such as LinkedIn, Monster.com, and chroniclevitae.com. Because of their limited coverage, however, these sources may be suitable primarily for supplementing the survey data rather than serving as a source of information for the sampling frame.

Summary. NCSES is casting a wide net in reviewing nonsurvey data sources that could be used in the development of sampling frames for the science and engineering workforce surveys. None of these data sources stands out currently as potential improvements over the sampling frames used for the NSCG or SDR. Continuing to monitor all of these data sources is important, however, because the availability, accessibility, characteristics, and relative benefits of these types of data are constantly changing.

Institution-Based Sampling Frame for the NSCG

As part of the GSS data collection, NCSES conducts a large-scale outreach effort annually to all U.S. academic institutions that grant research-based master's degrees or doctorates in science, engineering, and selected health fields. Although the data collected are at the aggregate level and are limited to students enrolled in master's- or higher-level programs, it may be possible to collaborate with academic institutions to obtain information for the NSCG sampling frame.

Indeed, between 1973 and 2010, the NCSES portfolio of science and engineering data collections included an additional survey, the NSRCG. Prior to the NSCG's transition to the use of the ACS for its sampling frame, the NSRCG was the primary source of new graduates in the NSCG sampling frame. The NSRCG was based on a two-stage sample design. In the first stage, a sample of academic institutions was selected using the list of institutions in the Integrated Postsecondary Education Data System database maintained by the National Center for Education Statistics. In the second stage, a sample of graduates was selected from the lists provided by the sampled institutions.

An institution-based sample for the NSCG would have several limitations and challenges compared with the ACS. The sampling frame would have to be supplemented with data from other sources in addition to the lists obtained from the academic institutions. A list of recent graduates obtained from institutions would not include people who joined the workforce many years ago or those who earned their degrees outside the United States but are now living here. As discussed, the options for developing a frame of individuals who obtained their degrees abroad may be particularly limited. NCSES is highly successful in obtaining data from institutions, but some government agencies are experiencing increasing difficulties with these types of requests, and this trend could have an adverse impact on response rates. In addition, it is important to note that the NSRCG faced some additional response rate challenges, primarily because of the difficulties associated with locating sample members who have left school and tended to be highly mobile during the years following their graduation. As the longitudinal aspects of the science and engineering surveys become more of a priority, however, some of these disadvantages may diminish somewhat if respondents are kept in the sample longer. Furthermore, educational institutions are increasingly motivated to track outcomes for their graduates, and this trend would be worth monitoring, possibly working closely with a small number of institutions.

Household-Based Sampling Frame for the NSCG

Although the focus of the NSCG is on college graduates who are in the science and engineering workforce, the target population is fairly broadly defined. The proportion of households with a member who holds a bachelor's degree in a science or engineering field or works in one of these fields is likely reasonably high (based on 2011–2015 ACS data, approximately 38.8 percent of households include at least one person with a bachelor's degree or higher). Although not likely to be more efficient than the current design, a household-based survey could potentially be considered as an alternative to the current NSCG design, particularly if the data collection could be combined with another survey. Another household survey also could potentially serve as a source of data for part of the sampling frame.

It might be possible to imagine the NSCG as a supplement to the Current Population Survey (CPS), which is a monthly rotating panel survey. In its current form, the CPS sample is not large enough to meet the NSCG's precision requirements, and it is unlikely to be capable of readily accommodating the increased respondent burden. If necessary, however, NSCG data could potentially be collected as a one-time, special module best administered by personal interview.

The cases leaving the CPS could be accumulated over time to provide a frame for a survey such as the NSCG. The monthly CPS sample size is approximately 60,000 households, and each household contains on average two working-age people (aged 15+). In other words, the CPS sample in a given month includes 120,000 individuals. One-eighth of the sample is replaced monthly, which means approximately 15,000 people leave the sample each month. Assuming that it would be permissible for the NSCG to survey people in the initial sample even before they left the CPS, a sample of 285,000 cases ($120{,}000 + 11 \times 15{,}000 = 285{,}000$) could be available in the first year. In subsequent years, an additional sample of 180,000 cases could be available ($12 \times 15{,}000 = 180{,}000$). Respondent burden, however, would be a concern for sample members who would be included in the follow-ups for both surveys. In addition, these numbers would not be sufficient to produce the level of detail currently published by NCSES. Even with the use of the ACS, which surveys 250,000 new addresses per month (compared with 7,500 new households in the CPS), it is necessary to aggregate multiple months to develop a frame for the NSCG (see Fecso et al., 2012, pp. 18–19).

Despite the challenges, the appeal of the CPS is that it could potentially serve as a source for a sampling frame of people who have been in the workforce for a while and obtained their degrees outside the United States. This sampling frame could then be used to complement a primarily institution-based sample of recent college graduates. In the long term, as the frame of bachelor's and master's graduates accumulated over years, the role of the CPS could be reduced. Eventually, the CPS could be used mainly or exclusively to sample people who received their degrees outside the United States.

Each of the potential frames discussed above has weaknesses, such as the lack of full coverage of the population, high cost, and incomplete data for stratification. At present, it appears likely that any alternative to the ACS as a sampling frame for the NSCG would have to involve the use of multiple frames.

RECOMMENDATION 4-5: It does not appear that any existing data sources would represent improvements over the current sampling frames for the National Survey of College Graduates or the Survey of Doctorate Recipients. Nonetheless, the National Center for Science and Engineering Statistics should continue evaluating alternative data sources for this purpose to ensure that it maintains awareness of the strengths and weaknesses of rapidly evolving options.

Other Sample Design Issues

Integrating the ECDS into the SDR

As discussed, the pilot ECDS relied on a sampling frame of individuals who earned their first doctorate or doctorate-equivalent degree within the past 10 years and who were working at an academic institution included in the GSS, in an FFRDC, or in the NIH IRP. The survey had a two-stage stratified sample design: after the institutional data collection was completed, individuals were selected from the institutions that responded. While the ECDS greatly expanded data collection on important emerging topics of interest, several challenges, associated primarily with the sample design, were experienced during the pilot study. First, the sampling frame is limited in its representation of employment sectors. Second, obtaining buy-in from the institutions was challenging, and NCSES has been experiencing increasing nonresponse among GSS institutions in general. Finally, demographic data for the second stage of the sampling design are lacking.

The question arises of whether the ECDS could be integrated more closely with the SDR. NCSES has considered administering the ECDS on a biennial schedule, alternating it with the SDR data collection. The possibility of a longitudinal component for the ECDS also has been considered, with follow-ups approximately 5 and 8 years after the initial survey cycle and the ECDS longitudinal respondents potentially being merged into the SDR sample when they age out of the ECDS target population. To implement a longitudinal design, however, NCSES would first have to overcome the response rate challenges that are currently impacting the ECDS.

As discussed in Chapter 3, some of the questions included in the ECDS would also be of interest for the SDR population. Accordingly, another possibility to consider is whether the SDR and ECDS samples could be integrated in a way that would treat the ECDS population as a subsample of the SDR. In this scenario, some questions would be administered to all sample members, and early career–focused questions would be treated as a module administered only to a subset of the respondents (see below and Chapter 3 for additional discussion of topic modules). If the sampling frame and response rates continued to be a challenge, the use of an SED-based sampling frame for early career doctorate recipients might have some advantages, although the sample would not include individuals with degrees obtained abroad.

RECOMMENDATION 4-6: The National Center for Science and Engineering Statistics should evaluate the feasibility of merging the Early Career Doctorates Survey into the Survey of Doctorate Recipients.

NSCG Primary Stratification Variable

The redesigned SDR sampling frame uses fine field of degree to stratify the sample. This approach is both important, because reporting by fine field of degree is of high value to the SDR, and straightforward, because the information is available on the sampling frame (SED). However, the NSCG is stratified by occupation. The question arises of whether the precision of key NSCG estimates could be improved by stratifying this survey as well by field of degree rather than occupation.

The panel does not believe that modifying the NSCG sampling frame to prioritize stratifying by field of degree would be beneficial. For degrees awarded at the baccalaureate level, and especially across the full age range of the workforce, an individual's current occupation is likely to be more informative than the degree obtained with respect to workforce capability, going beyond the science and engineering and nonscience and engineering classification. The SDR differs in this regard, because for doctoral degree holders, field of degree is likely much more relevant to workforce capability compared with baccalaureate holders.

Another question that arises is whether the surveys are collecting enough information, of the right kind, about individuals who hold a master's degree but not a doctorate. These individuals would appear to represent a vital component of the science and engineering workforce, and it might be useful to modify the design of the NSCG accordingly by oversampling those with a master's degree, at least among those with an undergraduate degree in a science and engineering field. It appears likely that a high proportion of individuals who hold a bachelor's degree in a science and engineering field and go on to obtain a master's degree obtain the latter degree in such a field as well.

RECOMMENDATION 4-7: The National Center for Science and Engineering Statistics should evaluate whether oversampling individuals with a master's degree who obtained a bachelor's degree in a science and engineering field would be a useful modification to the National Survey of College Graduates sample design.

Use of Matrix Sampling or Topic Modules in Questionnaire Design

A matrix sampling approach could provide one avenue for relieving the tension between needing some estimates reliable at the fine field level while not overburdening respondents with questions that need not be analyzed at that level. For example, the questionnaire items that are critical to fine field of degree could be administered to everyone, and items used only for estimates at higher levels of aggregation could be administered only to a subsample of respondents. This approach might also address NCSES's expressed need for a "quick turnaround" capability for some survey topics. An alternative design would be a short survey supplemented by administrative data and a long survey for a random subsample. The short survey could have more equivalent rates of sampling across fine fields. The short survey combined with the administrative data could be focused on meeting key reporting requirements, while the long survey could serve additional analytic purposes, including those of external data users. For sample members who did not consent to record linkage, the survey could default to the long version. These types of designs would be most feasible in electronic data collection modes and less realistic as long as a mail data collection mode was deemed necessary.

True matrix sampling is difficult to implement (see National Academies of Sciences, Engineering, and Medicine, 2016b). This approach would require reliable predictions of response rates for each stratum by mode (mail, Web, telephone). The use of matrix sampling in a mail survey would be especially challenging because multiple versions of the questionnaires would need to be produced and mailed. This approach would therefore be more feasible with a Web or telephone survey. Matrix sampling also has disadvantages for secondary users of the data as it would result in datasets with planned missing data, increasing the complexity of analyses.

For additional discussion of the potential use of topic modules (without a true matrix sampling design), see Chapters 3 and 5.

Possible Reconsideration of Design Objectives

It might be worthwhile to reflect on whether, by basing the designs of the SDR and NSCG primarily on prescribed levels of precision at very fine levels, the surveys are being rendered inefficient for many of the purposes for which the data are used. The prescribed precision levels may be necessary if one wishes to obtain estimates at very fine levels, but this capability comes at the expense of precision when one moves to more consolidated levels. Also, in determining the sample size requirements, it may be that improvements in precision that are being, or could be, obtained in estimation are not being considered. That is, it appears likely that the sample size calculations assume that Horvitz-Thompson estimation is being used, whereas in fact there is considerable scope for variance reduction through the use of auxiliary information, at least some of which is already being utilized.

The current designs have highly unequal sampling rates across different strata. This wide variation is likely to lead to considerable inefficiency for those estimates that cross stratum boundaries, as undoubtedly is the case for many of the key estimates. In the current NSCG sample design, a sample of 2,700 mathematicians represents a population of 200,000, while a sample of 5,000 represents the population of 3.3 million in science and engineering–related nonhealth occupations, resulting in average weights that are nine times higher than that of the mathematicians. In the case of the SDR, every graduate in behavioral analysis as well as health and medical psychology is included, but only 5 percent of clinical psychology graduates are included. This allocation makes sense if one is interested simply in data broken out by fine field of degree, but it is not a good allocation if one has any interest in those with a doctorate in psychology more generally, because clinical psychology graduates outnumber those other two groups combined by about 236 to 1.

Another disadvantage of the differential sampling rates that arise as a consequence of setting precision goals at the level of fine field of degree is that misclassification of a sampled unit can render the estimates almost meaningless. For example, if one of the 2,700 clinical psychology graduates sampled in the SDR actually obtained a degree in behavioral analysis, that one respondent would contribute 16 percent of the total estimate for behavioral analysis graduates. The question that arises is whether it is possible to achieve a better balance of the estimation goals for the SDR and NSCG at different levels. One enhancement that could potentially permit designs with less extreme differential sampling fractions yet maintain good precision by fine field of degree is the use of small area estimation techniques, discussed further below (see Recommendation 4-9).

ESTIMATION METHODS

The estimation procedures used for the NCSES workforce surveys are appropriate and well developed, but there may be opportunities to enhance the usefulness and reliability of estimates by using more sophisticated estimation methods than are currently employed. This section describes some avenues that could be explored to evaluate whether worthwhile enhancements to the current estimation methods are possible. The NSCG is used illustratively, but some of the ideas discussed are likely to be relevant to the SDR as well.

Composite Estimation

Composite estimation methods take advantage of the fact that respondents are included in a sample for several waves of data collection, and their responses to many questions are highly correlated over time. Because these conditions apply in the case of the current design of the NSCG, there is considerable scope for applying such a technique in deriving estimates for each cycle. Over time, methods of composite estimation have become increasingly sophisticated, and it might be profitable to invest resources in researching different options for use of such methods. One possible approach is discussed below.

If we are interested in estimating a population total Y_t for time t, then we denote its usual estimator by $\hat{Y}_t$. We define a different estimator, the K-composite estimator $\hat{Y}^C_t$, by considering not only the sample at time t but also the sample on the previous occasion, $t - 1$. The sample at time t consists of units (people in the case of the NSCG) that were also in the sample at time $t - 1$, which is denoted by the subscript $common_{t,t-1}$, and units that are new to the sample. Using only the common units, we can produce estimates of both Y_t and Y_{t-1}, denoted by

$$\hat{Y}_{common_{t,\, t-1},t} \text{ and } \hat{Y}_{common_{t,\, t-1},t-1}$$

respectively. Then, of course,

$$\hat{Y}_{common_{t,\, t-1},t} - \hat{Y}_{common_{t,\, t-1},t-1}$$

is an estimate of the change in Y, and it can be added to the previous estimate $\hat{Y}^C_{t-1}$ for time $t - 1$ to produce an alternative estimate for Y_t, namely

$$\hat{Y}^C_{t-1} + \hat{Y}_{common\ t,\ t-1,t} - \hat{Y}_{common\ t,\ t-1,t-1}.$$

We emphasize that

$$\hat{Y}_{common\ t,\ t-1,t} \text{ and } \hat{Y}_{common\ t,\ t-1,t-1}$$

are based on the same sample units. Because we now have two estimates of Y_t, it is natural to take a weighted average of the two:

$$\hat{Y}^C_t = (1 - K)\ \hat{Y}_t + K(\hat{Y}^C_{t-1} + \hat{Y}_{common\ t,\ t-1,t} - \hat{Y}_{common\ t,\ t-1,t-1}).$$

Because $\hat{Y}^C_t$ uses more information than $\hat{Y}_t$, it will often be a better estimator of Y_t.

The K-composite estimator was first used by the CPS but has since been replaced by more sophisticated forms of the estimator (U.S. Census Bureau, 2006, Ch. 10). The K-composite estimator can be used as a starting point, and if it is found to be useful for the NSCG, other types of composite estimation can also be considered. Applying the concept of composite estimation to the NSCG should be relatively straightforward since the sample components (rotation panels) are clearly defined. Developing a suitable variance estimator will require some thought. Note that both composite estimation and estimation for longitudinal analysis use survey responses spanning multiple time periods, but the goal of composite estimation is to produce *cross-sectional* estimates (in the case of the NSCG, estimates for the latest year of the survey).

RECOMMENDATION 4-8: The National Center for Science and Engineering Statistic should evaluate whether composite estimation techniques could be used to improve the National Survey of College Graduates and possibly the Survey of Doctorate Recipients estimates.

Model-Based Methods: Small Area Estimation

Because of the interest in highly detailed estimates (for example, by occupation, by field of study, and for various subpopulations), it is worth exploring the use of small domain estimation, more commonly referred to as small area estimation, techniques (Rao and Molina, 2015) for the NCSES workforce surveys. Using occupation as an example, it may be possible to group certain occupations into sets, where the behavior of units in each set is thought to be similar (with respect to variables of interest). If this grouping effort is successful, it will be possible to use small area estimation methods to "borrow strength" within a set, allowing better estimates for occupations (or for occupation by domain, such as race and gender) that have small sample sizes. In fact, if this approach is successful, it should be possible to reduce the sample size in some strata and either allocate the resources to other parts of the sample or invest the savings in data collection (e.g., better tracing in subsequent waves).

A good starting point would be to try the basic area-level model. Let y be the variable of interest (e.g., average income for different occupations). Then for occupation i, the survey estimator y_i is assumed to follow the model

$$y_i = \theta_i + e_i,$$

where θ_i is the true average, and e_i is the sampling error. The model assumes the existence of auxiliary variables that are related to income via the model

$$\theta_i = x_i'\beta + \varepsilon_i.$$

The "borrowing of strength" occurs via the parameter β, which will be estimated using sample data from all the occupations in a set. The literature contains numerous approaches to applying this model, with the greatest challenge being the need to estimate the variances of e_i and ε_i. The model can be generalized in different ways, including by adding the time element (the parameter of interest becomes θ_{it}). If suitable auxiliary information is available, it is also possible to consider using a unit-level rather than an area-level model.

If the research centers of excellence discussed in Chapter 6 were established, the potential uses of small area estimation techniques could be explored through that mechanism.

RECOMMENDATION 4-9: The National Center for Science and Engineering Statistics (NCSES) should evaluate whether small area estimation techniques could be used to enable reducing the sample size in some strata, thereby reducing the variability of the sampling rates used across strata and also perhaps freeing resources for use toward other NCSES objectives.

5

Data Collection and Processing

As described in Chapter 2, the Survey of Graduate Students and Postdoctorates in Science and Engineering (GSS) and Survey of Earned Doctorates (SED) are censuses, with a high degree of involvement from the relevant educational institutions. Over time, data collection procedures have been developed that result in very high response rates. The GSS collects data from approximately 700 academic institutions annually. Respondents are institutional coordinators who are asked to provide aggregate data on graduate students enrolled in science, engineering, or selected health field organizational units (such as academic departments and degree-granting programs) within the institution. The data collection consists of two stages. In the first stage, the list of eligible units at the institution is verified, and in the second stage, the aggregate student data are collected. Coordinators who respond on behalf of the institutions are contacted by e-mail, telephone, and FedEx letter, and they submit the data by Web or data upload or by sending an Excel file. Although the data collection is a substantial undertaking, the GSS has a response rate of more than 99 percent.

The SED is an annual census of individuals receiving doctoral degrees in that year. Unlike the GSS, it collects individual-level data directly from research doctorate recipients, and the initial request for participation comes from the educational institution at the time the degree is granted. This survey provides information on trends in degrees, the characteristics of the doctoral population, and postgraduation plans at the time of graduation. It also serves as an ideal sampling frame for the Survey of Doctorate Recipients (SDR). In administering the SED, NCSES receives substantial help from institutional coordinators because the data collected through the survey are valuable to the institutions. Asking students to complete the survey as part of the final stages of their graduation also is particularly effective because it is integrated into the graduation process, which marks an important milestone in the students' lives.

NCSES's data collection contractor follows up by e-mail, postal letter, postcards, and telephone with doctoral degree recipients who do not respond. Although respondents can participate in the SED by Web, mail, or telephone, NCSES has successfully transitioned the survey from a primarily paper-based data collection (whereby students returned completed mail questionnaires to the institutional coordinators, which were then forwarded to the data collection contractor) to a primarily Web-based survey (whereby responses are transmitted directly to the data collection contractor). Given the synergy of interests in having these data available and the resulting assistance from institutional coordinators, the response rates for the SED are typically 90 percent or higher. Although maintaining these response rates is becoming increasingly difficult as part of a trend that affects all surveys, the data collection procedures appear well suited to maximizing response rates efficiently.

This chapter focuses primarily on the National Survey of College Graduates (NSCG), the SDR, and the Early Career Doctorates Survey (ECDS). These three sample-based surveys have the most complex data collection procedures and face increasing implementation challenges at a time when the composition of the science and engineering workforce is undergoing rapid changes—in other words, at a time when the ability to produce high-quality measures of employment characteristics is more important than ever. These changes in the composition of the workforce also mean that the composition of the survey samples may be changing in ways that warrant adjustments to the data collection procedures.

OVERVIEW OF DATA COLLECTION PROCEDURES FOR THE WORKFORCE SURVEYS

NSCG

As discussed in Chapter 4, the NSCG has a rotating panel design that uses the American Community Survey (ACS) as a sampling frame for new cases. NSCG sample members are contacted by letter, by postcard, by telephone, and, in the case of sample members that have provided an e-mail address during a previous round of the survey, by e-mail (see Figure 5-1 for the contact strategy). Sample members who are characterized as having a low propensity to respond and a large sample weight are offered a $30 prepaid incentive.

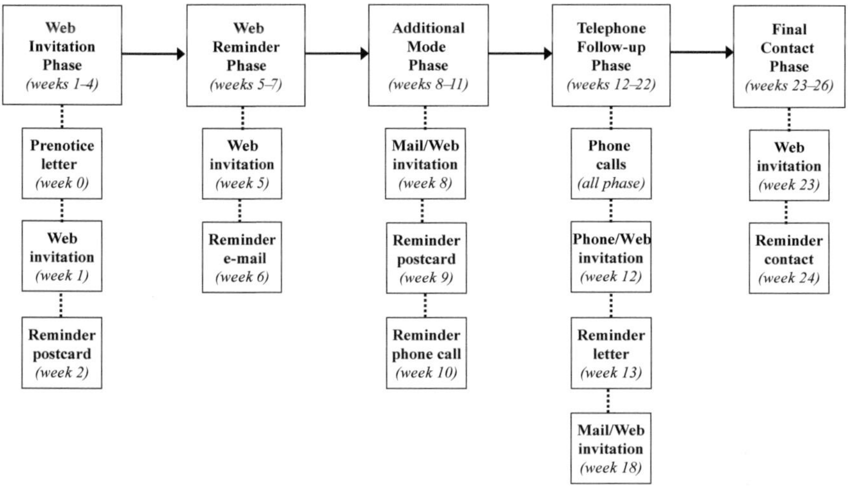

FIGURE 5-1 Sample member contact strategy for the National Survey of College Graduates.
SOURCE: Finamore (2016).

In 2015, the overall NSCG response rate was about 70 percent, slightly lower than the 74 percent response rate in 2013. This survey also was transitioned successfully to a "Web first" strategy, with mail and telephone response modes offered later in the data collection. This change increased the percentage of responses received by Web from 35 percent in 2010 to 77 percent in 2015 at considerable cost savings for the overall data collection (Finamore and Dillman, 2013). The remaining responses included 15 percent by mail and 8 percent by telephone. Although the great majority of responses are now received by Web, mail and telephone remain important modes of contact with sample members, and indeed the initial Web invitation is sent by postal letter (see below for detailed discussion).

As is the case for any longitudinal survey, one challenge for the NSCG is panel attrition. Because differential attrition and its impact on data quality are of particular concern, the NSCG design limits the number of follow-ups to four rounds of biennial data collection. In other words, sample members remain in the survey for three rounds after the initial survey cycle. Since 2010, to minimize attrition bias, sample members who participated in the initial interview have been considered eligible for follow-up, regardless of the case's disposition in other, previous follow-ups. In the 2015 NSCG, this approach resulted in a 53 percent unconditional response rate for the sample obtained from the 2009 ACS. The conditional response rate was 80 percent among those who had responded in the previous round and 26 percent among those who had not responded in previous rounds.

SDR

The SDR survey sample is based on the SED census. Most SDR sample members are contacted approximately 2–3 years after their participation in the SED, and they receive follow-ups every 2 years until they reach the age of 76 or otherwise become ineligible. Sample members are contacted by letter, postcard, e-mail, and telephone. Figure 5-2 shows the SDR contact strategy. Over the years, NCSES has continued to refine the incentive strategy for the SDR. The survey currently uses a prepaid incentive of $30, strategically targeted to reduce nonresponse bias depending on the sampling group and whether an incentive was needed for the sample member in previous rounds of data collection. The survey achieves a response rate

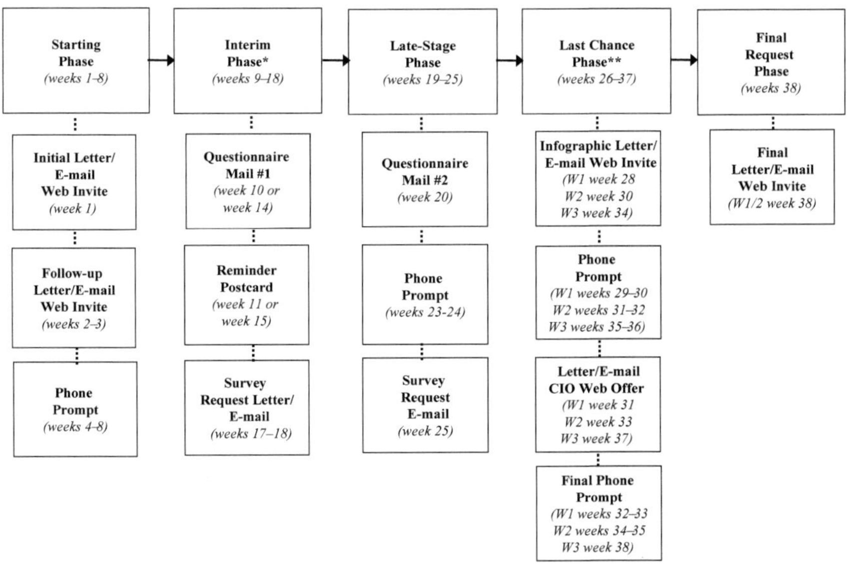

FIGURE 5-2 Sample member contact strategy for the Survey of Doctorate Recipients.
*There was a partial contacting hiatus during weeks 5–13.
**To allow adequate locating of newly selected sample member, the Last Chance contact sequence was deployed in three waves as cases continued to be found.
SOURCE: Finamore (2016).

of about 63 percent. In 2015, 82 percent of the respondents completed the survey by Web, 10 percent by mail, and 8 percent by telephone.

Nonresponse in the SDR is associated primarily with challenges of locating the sampled individuals rather than with lack of willingness to participate. Sample members with incomplete information in the SED sampling frame and noncitizens tend to be particularly difficult to locate. To increase the proportion of cases that are successfully located, the NCSES data collection contractor performs several locating activities prior to the beginning of the field period (and prior to contact with sample members).

ECDS

As discussed in Chapter 2, the ECDS is only now moving to full implementation, following completion of a pilot test that NCSES judged to be successful. The survey has a two-stage sample design that involves a stratified probability sample of institutions in the first stage and a stratified probability sample of individuals in the second stage. For the pilot, a combination of mail, e-mail, and telephone contacts was used for both the institutional contacts and individual sample members. Institutional coordinators were asked to submit lists of eligible employees electronically, and NCSES's data collection contractor drew from these lists samples of individuals to be contacted. Pretesting showed that having someone from an institution reach out to individual sample members led to higher compliance relative to initial contact from the National Science Foundation, so institutional coordinators were asked to send an initial e-mail introducing the survey. Following the initial contact, NCSES's data collection contractor managed the subsequent follow-ups. The response rate for the pilot study was 56 percent. Sample members could respond by Web or telephone, and the Web was used by nearly all respondents (99.5%) during the pilot test. Figure 5-3 shows the sequence of contacts with individual sample members for the ECDS.

Table 5-1 summarizes the key data collection characteristics of the NCSES education and workforce surveys. The remainder of this chapter focuses on aspects of the data collection and processing that would particularly benefit from changes or enhancements.

DATA COLLECTION MODES

NCSES has been actively researching the optimal use of different data collection modes for its science and engineering workforce surveys. The question of how to optimize the use of different modes to increase contact, encourage cooperation, enhance data quality, and reduce costs is complex and dependent on the target population and types of questions asked.

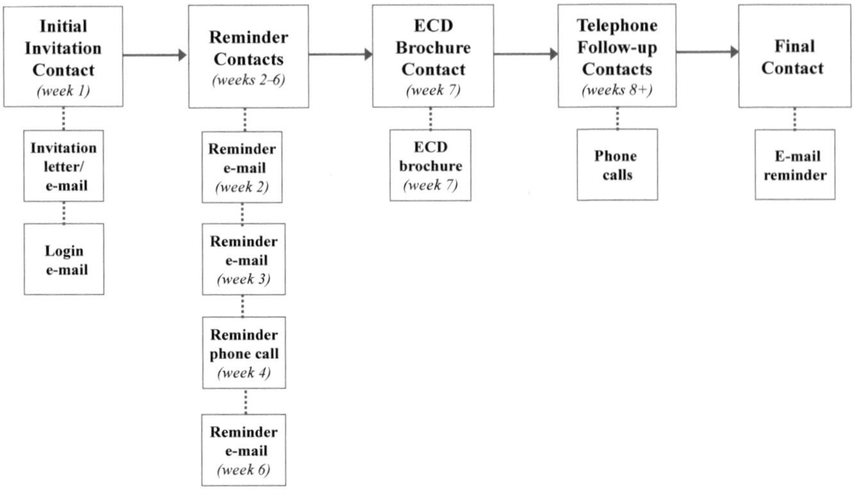

FIGURE 5-3 Sample member contact strategy for the Early Career Doctorates Survey.
SOURCE: Finamore (2016).

Ongoing research is needed in this area for a number of reasons. First, there is an intricate interdependence among the different modes of contact and different response options. Even if Web responses are the goal or the only response mode offered, contact through another mode, such as mail or telephone, often is necessary, and combining modes in this way can substantially boost response rates. Mail contact, such as an advance letter, also can serve the purpose of legitimizing the survey and delivering incentives to respond. Second, from a respondent's perspective, the relative importance and desirability of different modes of communication and response are constantly evolving. It is increasingly more difficult to reach sample members by voice telephone, while the role of electronic communication, particularly by smartphone, is increasing. Third, although NCSES's definitions of the target populations for its surveys have not changed substantially in recent years, the characteristics of the samples are constantly evolving. For example, the increase in sample members residing abroad presents new challenges in terms of locating and communicating with them.

Consequently, one area of research that deserves further consideration is the optimal mix of modes of contact and modes of response that are offered or encouraged. A related issue is the optimal use of incentives. NCSES has been evaluating these issues with particular emphasis on the NSCG, which must begin with contact by mail in the first year of data collection because other forms of contact information from the ACS sampling frame are more limited. At the end of the NSCG, respondents are asked to provide telephone and email information, which enables follow-ups through multiple modes in subsequent waves. The use of multiple contact modes improves response rates because the survey reaches a larger proportion of the sample members and because using a variety of modes to communicate with sample members offers multiple opportunities to deliver persuasive messages.

Although NSCG sample members are contacted initially by mail, they are asked to complete the survey by Web, and a mail questionnaire generally is not provided until later stages in the follow-up process. As discussed above, this "Web first" data collection strategy, tested in 2010 and fully implemented starting with the 2013 NSCG, substantially increased responses by Web. NCSES has found that the use of this approach maintains the overall response rate at a lower cost relative to either telephone or paper questionnaire first strategies (Finamore and Dillman, 2013).

One question that deserves consideration is whether it is worthwhile to continue providing a mail response option in the later stages of the data collection, or the mail contact could simply direct sample members to the Web after a paper questionnaire has been included once. A mail response option is expensive because the questionnaires must be printed, and the responses must then undergo data entry or scanning. Eliminating the paper questionnaires altogether also could greatly increase the flexibility of the survey instruments because electronic modes can more readily accommodate complex branching and the tailoring of questions on the basis of information available from prior rounds of data collection or from other data sources, such as administrative records. In addition, limiting the response options to Web and telephone would make matrix sampling options, or the use of topic modules, more feasible. However, the implications of eliminating the mail response mode would need to be carefully researched, particularly in light of the rapid decline in the acceptability of voice telephone as a response (and contact) mode, and because the mail contact mode would still be necessary even if a questionnaire were not mailed.

NCSES collected data and carried out experiments to better understand the role of the mail questionnaire, particularly in the NSCG. When respondents are asked about their response mode preferences at the end of their first NSCG questionnaire, somewhat fewer than one in five indicate that they would prefer to complete the survey via a questionnaire sent by mail. In the 2015 NSCG, approximately 15 percent of the returned sample responses were received by mail, but among those who indicated a preference for a mailed questionnaire in 2013 and completed the survey in 2015, just over half (54%) responded by mail (approximately 40% completed the survey by Web and 7% by telephone) (information provided by NCSES

TABLE 5-1 Key Data Collection Characteristics of the NCSES Education and Workforce Surveys

Survey	Target Population
Survey of Graduate Students and Postdoctorates in Science and Engineering (GSS)	Graduate students, postdoctoral appointees, and doctoral-level nonfaculty researchers at a U.S. academic institution granting research-based master's or doctoral degrees in science, engineering, and health fields (information collected from institutional coordinators)
Survey of Earned Doctorates (SED)	Recipients of research doctorate degrees in any field, graduating from U.S. academic institutions
National Survey of College Graduates (NSCG)	Individuals living in the United States who have at least a bachelor's degree in any field
Survey of Postdocs at Federally Funded Research and Development Centers (FFRDCs)	Postdocs employed at the FFRDCs
Survey of Doctorate Recipients (SDR)	Individuals with a science, engineering, or health research doctorate degree from a U.S. academic institution (sample based on the SED)
Early Career Doctorates Survey (ECDS)	Postdoctoral fellows and other doctoral degree recipients working at a U.S. academic institution, at an FFRDC, or in the National Institutes of Health Intramural Research Program who received a doctorate-equivalent degree in the past 10 years in any field and any country

at the panel's request). To potentially increase the efficiency of subsequent contacts, information collected from respondents when they first participate in the survey could be expanded to include questions about access and preferences for desktop computers, tablets, and smartphones, and to obtain permission for contact by text messages. However, the cost implications of further customizing modes of contact would have to be carefully researched, in particular in comparison with strategies that encourage participation by a certain mode (such as the Web-push strategy). Research on the impact of telephone contact could be focused on attempting to determine the type

Data Collection Frequency	Contact Modes	Response Modes	Response Rates (most recent year available)	Incentives
Annual	E-mail, telephone, letter	Web, data upload, Excel file	Overall: 99.5% Web: 89.0% Data upload: 9.8% Excel file: 1.1%	None
Annual	E-mail, letter, postcard, telephone	Web, mail, telephone	Overall: 90.2% Web: 93.0% Mail: 5.4% Telephone: 1.6%	None
Biennial	Letter, postcard, e-mail, telephone	Web, mail, telephone	Overall: 70% Web: 77% Mail: 15% Telephone: 8%	$30 prepaid for a subset of the respondents
Periodic (every 2–3 years)	Letter, e-mail, telephone	Web	100%	None
Biennial	Letter, postcard, e-mail, telephone	Web, mail, telephone	Overall: 63% Web: 82% Mail: 10% Telephone: 8%	$30 prepaid, depending on case
To be determined	Letter, e-mail, telephone	Web, telephone	Overall: 56% Web: 99.5% Telephone: 0.5%	To be determined

of phone the respondent has and to examine the role of caller IDs, leaving messages, and sending text messages as reminders.

The 2015 NSCG included an experiment that varied when the mail questionnaire was first provided to returning sample members. The four experimental conditions were as follows: (1) mail questionnaires in weeks 8 and 18 (control group), (2) mail questionnaire in week 8 and Web invitation in week 18, (3) Web invitation in week 8 and mail invitation in week 18, and (4) Web invitations in weeks 8 and 18. Respondents who had indicated a preference for a mail questionnaire in a previous round were randomly assigned to one of the four groups, but also received a mail questionnaire in week 1. The overall response rate across modes did not vary by experimental condition. However, there was a small difference in when people responded: the use of a paper questionnaire in week 8 rather than a Web invitation resulted in an incremental response rate increase of approximately 2 percentage points.

While this research provides valuable insight into the potential consequences of removing the mail response option, further research is needed on the role of the paper questionnaire. A mailed questionnaire appears to make a difference even if sample members choose to respond by a different mode. There are several possible explanations for this finding. For some respondents, the ability to see the questions ahead of time clarifies the purpose of and expectations for the study and so removes a barrier to completing the survey. This informative function of a paper instrument may be particularly important for sample members who are new to the survey. Another possible explanation is that a mailing including a paper questionnaire simply serves as an additional form of communication differing from previous ones, and boosts response rates in the same way as other strategies focused on varying the types of contact and incentives. It is possible that increased telephone contact could serve a similar function for at least a subset of respondents, but contacting sample members by telephone is becoming increasingly difficult and less cost efficient.

NCSES has a commendable record of conducting experiments to improve the response rates and control the costs of its surveys, complemented by publishing the survey results and making them accessible to the scientific community, including other statistical agencies facing similar challenges. Continued investigation into the interaction among mode of contact, mode of response, and the timing and amount of incentives will be needed as the characteristics of the sample change over time. Again, research conducted in this area by other government agencies, such as the Census Bureau's research for the 2020 census and ongoing research for the ACS, could provide useful insights on which to build.

RECOMMENDATION 5-1: The National Center for Science and Engineering Statistics (NCSES) should continue to research the optimal mix and sequence of contact, response modes, and incentives. Lessons learned from the National Survey of College Graduates should be evaluated in the context of the other surveys, particularly the Survey of Doctorate Recipients. NCSES should ensure that costs are tracked in sufficient detail, for both new and continuing participants, to enable sophisticated comparisons of the different options, along with the implications for data quality.

SURVEY CONTENT

The content of the NCSES surveys has undergone many changes over the years as the agency has attempted to keep the questionnaires current and anticipate data needs for the near future. As discussed in Chapter 3, there are emerging data needs that could be met by changes ranging from relatively modest additions to the questionnaires (e.g., adding some of the questions from the ECDS to the SDR, and possibly to the NSCG) to the development of modules that could be used to accommodate data collection on more complex topic areas (e.g., modules based on the ECDS content, such as professional activities or work–life balance, or a module on entrepreneurship and innovation). New questions or modules could be particularly useful if the content were coordinated across all of the surveys to which it is applicable, which is in line with how NCSES has always approached the development of new questions.

A major challenge for NCSES when considering questionnaire updates is that the surveys serve an essential role in monitoring trends. In addition to new data needs, however, the relatively rapid changes in the workforce discussed in this report necessitate continuous review and updating of the terminology used in the existing questions. As discussed in Chapter 3, for example, the question on training needs to be updated to prompt respondents to consider newer forms of training in their responses. Sources of funding for education represent another key area of interest that undergoes constant changes the questionnaires need to reflect. Some questions could simply use further refinement because they may be too technical or ask about circumstances with which many respondents may not be familiar. For example, the SED question on tuition remission asks respondents whether they received a "full or partial tuition remission (waiver)." It is not clear whether graduate students would know whether they had received an out-of-state tuition remission or whether they would consider any tuition they had not paid out of pocket as "waived."

Participants in the workshop held for this study (see Chapter 3) indicated that in its current form, the questionnaire item about the employer's location may become increasingly less informative as more people telework or work from locations other than where their employer is located. Figure 5-4 shows the current question wording, including the respondent instructions, which attempt to capture unconventional scenarios. NCSES has tested this approach, but it is not clear whether increasingly more complex working arrangements mean that fewer respondents are able to interpret the question as it was intended. It is also possible that tailored approaches using a series of simpler questions (e.g., "Did you have more than one job?", "Did your employer have more than one location?") rather than instructions might yield more reliable and valid answers.

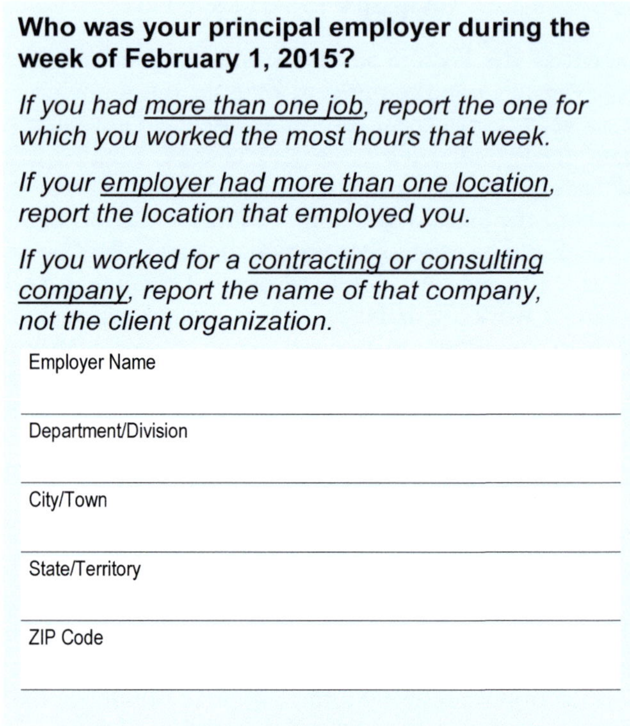

Who was your principal employer during the week of February 1, 2015?

If you had more than one job, report the one for which you worked the most hours that week.

If your employer had more than one location, report the location that employed you.

If you worked for a contracting or consulting company, report the name of that company, not the client organization.

Employer Name

Department/Division

City/Town

State/Territory

ZIP Code

FIGURE 5-4 Principal employer question on the National Survey of College Graduates and Survey of Doctorate Recipients.
SOURCE: National Survey of College Graduates Questionnaire. Available: https://www.nsf.gov/statistics/srvygrads/surveys/srvygrads-newrespond2015.pdf [December 6, 2017].

Keeping concepts and terminology current is an ongoing challenge for all longitudinal surveys, and there is no well-developed methodology for doing so. Nonetheless, periodic and regular reassessment of concepts and terminology is needed in any longitudinal study. Regular programs of cognitive interviewing, ongoing random probes (Redline, 2013; Schuman, 1966), and feedback from analysts who work with specialized subpopulations each could contribute to the process, as could the outside advisory mechanism discussed with and/or included in Recommendation 3-10 in Chapter 3.

New topics also emerge as times change. When NCSES identifies the need for new content areas, it needs to continue to carry out a full development program that includes a literature review, input from stakeholders, question development and testing, and usability testing for the instrument across all modes. New items need to be considered for inclusion in all surveys for which they are relevant, either as core questions or as topical modules, to promote comparability of data across surveys. The potential impact of any new items on trend data needs to be researched and tested with particular care, and possibly benchmarked when new items are introduced.

RECOMMENDATION 5-2: The National Center for Science and Engineering Statistics should continue to monitor, and formally evaluate as needed, the content of its survey questionnaires to ensure that the concepts and terminology are up to date and familiar to respondents. Changes should be implemented with careful consideration of their impact on trend data.

Topic Modules

As discussed in Chapters 3 and 4, increased reliance on topic modules is a potential solution that could reconcile the need to provide fine-grained data on core questions with the need to gather detailed data periodically for an in-depth understanding of certain topics. To a limited extent, NCSES already relies on questionnaire structures that include core topics and occasional modules, but the modules are essentially special topics that are administered to the entire sample. At the other end of the spectrum, as discussed in Chapter 4, true matrix sampling is difficult to implement and produces datasets that are difficult to use. Looking forward, it may be possible to move closer to a middle-ground approach, particularly if the surveys become less dependent on paper-and-pencil questionnaires (see below) and if greater integration can be achieved with administrative records databases that could trigger custom surveys for subsets of the sample.

From an implementation perspective, the first round of data collection for each sample member would have to be a comprehensive survey, likely without modules. Some of the data collected during this initial round could then be used to target modules in subsequent rounds. For example, those working in the private sector could receive questions about training, skills, and experiences specific to this sector. As discussed in Chapter 4, it might also be possible to integrate the ECDS with the SDR and administer to early career sample members modules applicable to them. This approach would require incorporating a career life-cycle perspective into the selection of the survey content and modules. Samples for various modules could be staggered in replicates to smooth fieldwork (possibly moving to year-round data collection) and management. With a design of this type, early replicates could be used to obtain data that would in turn be used to improve estimates and fielding of later replicates, as is done for the National Survey of Family Growth (Centers for Disease Control and Prevention, 2015).

As discussed in Chapter 3, the use of topic modules may increase not only the complexity of the survey but also response burden. Greatly increasing the length of the survey instruments may not be feasible in light of the need to adapt them for smartphone administration (see below). A series of shorter follow-up surveys could be considered instead of periodic longer surveys, with intervals determined by the stage in a person's career or potentially information from administrative records. This approach could increase costs and could also adversely impact participation rates. If the surveys were viewed as true panel surveys, however, it might be possible to further refine contact and incentive strategies that would facilitate more frequent follow-ups. Techniques to this end could include obtaining more detailed contact information (such as multiple e-mail addresses) and more information about available technologies (desktop, tablet, smartphone) and preferred modes of contact. Sample members also could be asked about the likelihood of their moving or changing jobs in the near future.

An additional consideration for the use of modules is that if the questionnaires became highly customized, resulting context effects could have an impact on the trend data available for the other items, and potentially the core items, on the surveys.

Government surveys exist that use topic modules in a variety of different ways, and the experiences of other agencies could provide valuable insights to NCSES. Some of the surveys that could be reviewed are the National Health Interview Survey, the Survey of Income and Program Participation, and the Current Population Survey.

RECOMMENDATION 5-3: Looking ahead, the National Center for Science and Engineering Statistics (NCSES) should explore placing greater reliance on topic modules to address the need for occasional detailed data on selected topics for subgroups. NCSES should evaluate whether the challenges associated with implementing topic modules might be reduced and the benefits increased as a result of the recent increased reliance on the Web mode, as well as potential integration with data from administrative records and the development of tailored field procedures to reduce respondent burden (see Recommendation 3-2 in Chapter 3).

Optimizing the Survey Instruments for Smartphones

One area that deserves attention going forward is the increasing use of mobile devices to access Websites, including Web surveys, because this trend has challenging implications for the design and content of the NCSES survey questionnaires. In the 2015 NSCG, approximately 4 percent of Web surveys were completed by smartphone and an additional 8 percent by tablet computer (see Table 5-2). These numbers, and particularly respondents' expectations for seamless interaction with a Web instrument on a smartphone, are likely to increase over the coming years. As new modes of data collection have been introduced, NCSES has worked on redesigning its questionnaires to reduce measurement error and minimize potential mode effects across the mail, voice telephone, and Web modes. For example, many of the long "check-all that apply" lists have been replaced with "yes/no" questions as a result of research supported by NCSES (Smyth et al., 2006, 2008). NCSES also undertook a research project to optimize the 2017 NSCG instrument for mobile devices. If the server detects that a mobile device is being used for the NSCG, the layout will adjust to provide larger font, buttons, and spacing and to eliminate the need for horizontal scrolling. Additional research could focus on revising remaining questions with relatively long lists of answer options and on turning definitions and instructions into questions (e.g., asking how many classes of a particular type people took instead of instructing them to exclude such classes) as part of an overall effort to unify the instrument construction across modes.

The challenge of adapting complex surveys to smartphones is not unique to NCSES, and best practices for meeting this challenge have not yet been widely developed and accepted in the survey industry. Research in this area is ongoing, and NCSES needs to continue to monitor new developments and conduct research on potential advancements. This research needs to be expanded to all of the agency's science and engineering workforce surveys.

TABLE 5-2 National Survey of College Graduates Web Survey Response by Device, 2013 and 2015

2013 NSCG		Number of Web Respondents	Mean Completion Time	Median Completion Time
Device	Smartphone	766	35.8	32.1
	Tablet	5,595	30.0	26.2
	Desktop/Laptop	56,021	27.8	24.2
2015 NSCG		**Number of Web Respondents**	**Mean Completion Time**	**Median Completion Time**
Device	Smartphone	2,123	31.0	27.7
	Tablet	4,644	28.3	25.4
	Desktop/Laptop	52,676	25.8	22.5

SOURCE: Fiegener and Finamore (2016).

RECOMMENDATION 5-4: The National Center for Science and Engineering Statistics should continue to conduct research on ways of optimizing its surveys for smartphone administration and to monitor further developments in questionnaire administration and response technology. This research should be conducted toward the overall goal of unifying the instrument construction across the different survey modes and, to the extent applicable, the different surveys.

MAINTAINING PANEL PARTICIPATION AND RESPONSE RATES

NCSES has a number of completed and ongoing research projects focused on maintaining participation and response rates. In recent years, the agency also has begun implementing adaptive design, which enables researchers to tailor data collection decisions in ways that optimize costs and data quality. NCSES began experimenting with adaptive design for the NSCG in 2013, and the 2015 mode experiments described above can be considered part of these efforts. NCSES found that active data monitoring and adaptive data collection interventions during the fieldwork improved representativeness. The agency is gradually building up research to inform possible interventions, such as targeted questionnaire mailing, special routing or handling of cases in the telephone mode and prioritization of calls, and changes to the default response mode for subsets of the cases. Adaptive design also was implemented in the SDR in 2015, with the goals of cost-efficiently minimizing response bias and maximizing the representativeness of key domains.

The use of paradata, such as the use of information from contact histories to tailor subsequent contacts, can prove especially useful in guiding decisions about follow-ups. The potential use of information from external data sources, such as the unemployment insurance earnings database, is another area that needs to be explored in the context of adaptive design, particularly for the NSCG (see Chapter 3).

NCSES's collaboration with the Census Bureau to assess the potential feasibility of using alternative data sources to expand the usefulness or efficiency of the NSCG also could be expanded to include the use of administrative data, such as information from the unemployment insurance earnings database, in the data processing stages to evaluate nonresponse bias or subsequently to contribute to nonresponse weights. The possibility of using administrative data in the processing stages of the SDR needs to be considered as well. Although the Census Bureau may not be able to share individual-level data with NCSES for this purpose, particularly for the SDR, which is not conducted by the Census Bureau, a "blinded" process could be developed to obtain group characteristics for batches of sample members, based on limited identifying information sent to the Census Bureau.

Experiments conducted in the NSCG revealed that offering incentives early is the approach that best balances costs, response rates, and the representativeness of the sample. NCSES also found that offering incentives to past incentive recipients increases response rates and representativeness. The agency continues to experiment with and refine the optimal use of incentives, and this research could be expanded to evaluate whether certain types or timing of incentives can further increase Web response among subgroups of respondents.

Research in the area of adaptive design is still in the early stages, but the development of strategies that increase the flexibility of large government data collections such as the NSCG and the SDR has promising implications for increasing efficiency. This type of research is particularly useful in balancing considerations related to increasing participation, reducing nonresponse, and controlling costs. Publishing research on these topics in methodological journals also is important to expand knowledge in the area in general and enable others to build on NCSES's work.

As discussed, the NCSES surveys face particular nonresponse challenges due to panel attrition, which has at least two components: those who are difficult to locate and those who may not be motivated to participate. In the case of both the NSCG and SDR, these two causes contribute about equally to nonresponse, although there are differences among subgroups.

While NCSES has worked continuously to build on the latest research in the field toward improving response rates overall, the understanding that could be gained from the experience of other longitudinal studies in particular may be underutilized. The repertoire of techniques used to maintain high response rates for several large-scale longitudinal government surveys could be considered. Although such techniques as sending newsletters with study results involve expense, these contacts without a request can communicate to sample members that their contributions are appreciated and cultivate a sense of ownership. These techniques also can help with keeping address lists current, which is particularly important for keeping in touch with sample members who move overseas.

RECOMMENDATION 5-5: The National Center for Science and Engineering Statistics should continue research aimed at expanding the use of adaptive design in the National Survey of College Graduates and Survey of Doctorate Recipients. Experiments involving contact schedules, incentives, and modes should focus on the goals of increasing participation, reducing nonresponse bias, and controlling costs.

RECOMMENDATION 5-6: The National Center for Science and Engineering Statistics should evaluate the feasibility of using administrative data, such as unemployment insurance earnings data, to inform adaptive

design, address nonresponse due to lack of contact information, conduct nonresponse bias analysis, and develop nonresponse weights.

RECOMMENDATION 5-7: The National Center for Science and Engineering Statistics (NCSES) should consult with other agencies that have longitudinal government surveys in their portfolios to identify techniques they have used successfully to increase the ability to locate sample members, motivate continued participation, and prevent panel attrition in the surveys. NCSES should develop an integrated approach to building and maintaining relationships with panel members and evaluate its success. Insights from research on these strategies should then be shared with other agencies.

DATA PROCESSING

One of the desires voiced most frequently by data users during the workshop held for this study was for more timely data. Given the increase in Web responses and continuous improvements in data cleaning and data entry procedures in general, it may be worthwhile to evaluate the data processing procedures for the science and engineering workforce surveys to assess (1) whether all types of errors that are found and corrected are sufficient in number to warrant the time and expense, and (2) whether any steps in the process lead to delays in public release of the data. Steps that do not result in quality improvement (e.g., cleaning of data from Web surveys) might then be reduced, and steps that lead to delays in data release might require redesign. Research conducted in the area of establishment surveys could provide a perspective that NCSES has not yet explored (see Biemer and Lyberg, 2003; Smith et al., 2003).

RECOMMENDATION 5-8: The National Center for Science and Engineering Statistics should evaluate the data processing methods used for the surveys to identify potential opportunities to increase efficiency and the timeliness of data releases.

6

Data Analysis, Dissemination, and Outreach

This chapter reviews opportunities to enhance data analysis and dissemination efforts and increase outreach to stakeholders. The observations herein are based on input provided during the data user workshop held for this study and throughout the panel's work, as well as on the panel members' own experience. It is important to note that NCSES has always actively disseminated information about its surveys, data, and data products, and continues to do so despite staff resource limitations. For example, the agency recently published an Info Brief on expansion of the data from the Survey of Doctorate Recipients (SDR) (National Center for Science and Engineering Statistics, 2017a), a potentially impactful way to distribute some of the findings based on the new data, as well as discuss the new methodology. The use of Webinars to disseminate information about new reports, such as the recent Webinar on the *Women, Minorities, and Persons with Disabilities in Science and Engineering* (*WMPD*) report, is another example of the agency's useful outreach efforts. As mentioned in Chapter 3, NCSES also recently launched a project focused on further expanding stakeholder outreach.

The panel acknowledges the staff resource limitations that make it difficult for NCSES to increase its emphasis on additional analysis and dissemination. The recommendations in this chapter focus primarily on strategies that could leverage existing mechanisms and build collaborations that could grow and strengthen the surveys' stakeholder base with relatively low resource investments.

KEY DATA PRODUCTS

To fulfill its mandate of providing information about the state of science and engineering in the United States, NCSES produces in a typical year more than 30 reports based on its science and engineering workforce surveys. These include two flagship publications that are congressionally mandated: the *Science and Engineering Indicators* (*SEI*) and *WMPD* reports. The *SEI* report is prepared under the guidance of the National Science Board (NSB) and is updated biennially with new data to inform policy decisions. *WMPD,* a report prepared by NCSES and mandated by the Science and Engineering Equal Opportunities Act, is focused on the participation of women, minorities, and persons with disabilities in science and engineering education and employment. The *WMPD* report also is updated biennially.

In addition, NCSES issues companion pieces to these two main reports. For example, there are *Digests* for the *SEI* and *WMPD* reports, which are condensed versions of the longer reports and highlight some of the most important indicators. *Revisiting the STEM Workforce*, NSB's companion piece to the 2014 *SEI* report, presents 2014 *SEI* data in the context of policy questions and illustrates ways in which the data can inform policy debates. In addition to reports, NCSES publishes shorter information briefs that summarize issues and trends based on the data. A variety of tables also are regularly published.

For researchers who wish to conduct their own analyses, the data are available in a variety of ways:

- public-use microdata files (available to download through the NCSES Website);
- public-use online data tools (e.g., WebCASPER, Scientists and Engineers Statistical Data System [SESTAT], SED Tabulation Engine);
- restricted-use license agreements, available for access to the National Survey of Recent College Graduates (NSRCG), SDR, SESTAT integrated file, Doctorate Records File/Survey of Earned Doctorates (SED); and
- Federal Statistical Research Data Centers (FSRDCs).

FSRDCs are managed by the Census Bureau and typically entail partnerships between federal statistical agencies and research institutions, such as universities. Currently, only National Survey of College Graduates (NSCG) data are available through FSRDCs, but NCSES is involved in discussions about expanding the use of FSRDCs for other NCSES surveys as well. A data enclave also is being tested.

As discussed in Chapter 3, NCSES has to balance requests from a variety of stakeholders when deciding on the content (and to some extent the design) of its surveys. The prioritization of resources invested in data dissemination and outreach also deserves careful consideration based on stakeholder needs, including sufficient attention to building new stakeholder groups.

From the perspective of NCSES's mandate, the flagship reports are a priority. Beyond those reports, additional efforts invested in analysis and broader dissemination of findings would represent opportunities for NCSES to increase awareness of the data and their impact, especially among policy makers. On the other hand, researchers have an increasing need for access to the microdata, and in particular the restricted-use data. The remainder of this chapter addresses two broad categories of stakeholder needs: (1) access to microdata, and (2) availability of publications, which for the purposes of this discussion includes preformatted data tables.

ACCESS TO MICRODATA

The NCSES science and engineering workforce surveys are of interest to a large group of stakeholders who would like to use the data for in-depth or customized analysis. Many of these stakeholders are academic researchers, but data users from many other sectors, such as nonprofit groups, professional societies, and the media, as well as international users, have this same need.

Historically, some of the data user needs could not be met even with access to the microdata files because of the small sample sizes in the public-use files. Given the types of highly specialized work being conducted by many scientists and engineers, some workshop participants expressed the need to be able to tabulate and model the data at a finer level than what has been available through the public-use files. For example, they would like to be able to separate doctorate recipients in nuclear engineering from other types of engineers to model the supply for clients such as the Department of Energy. They also would like to be able to conduct more detailed analyses by demographic group. For example, they would like to be able to answer questions for minority females, not just minorities and females separately (National Academy of Sciences et al., 2011; Pearson et al., 2015; Slaughter et al., 2015). In the case of the SDR, more of these data will become available as a result of the redesigned sample. NCSES is also in the process of reevaluating its disclosure control practices to determine whether more data could be included in the public-use microdata file in general.

Some existing research questions, including the need for more in-depth research on small population groups, such as racial or ethnic minorities, could be answered with the available restricted data but are not pursued because access to these data has been limited or cumbersome in the past. Increasing the number of datasets available through FSRDCs would alleviate this problem to some extent, but continued collaboration with other agencies also is important to streamline the processes for accessing the data so as to alleviate the burden associated with establishing data access for researchers. In particular, technological innovations are making it easier to enable researchers to conduct analyses of the restricted data in a secure environment, without the need to travel.

Researchers would find it useful if linkages to other datasets were easier. Such requests currently are evaluated on a case-by-case basis, and the linkage is carried out by NCSES or an NCSES contractor. This approach is time-consuming and also makes replication difficult. A possible way to streamline the process would be to support research centers that could assist with these tasks. Many universities already are involved with the FSRDCs, and indeed, approximately 90 percent of the restricted-use licenses are granted to academic researchers; thus it might be advantageous to locate research centers in academic institutions. The centers also could take the lead on analysis projects and conduct their own dissemination. Importantly, these types of centers would lead to the growth of research communities around the surveys, starting with faculty and graduate students who wanted to conduct research on these topics. It might be particularly useful to consider establishing such a center at a minority-serving institution, which could not only increase the use of the data for research related to underrepresented groups but also boost the traditionally lower response rates among these groups.

RECOMMENDATION 6-1: The National Center for Science and Engineering Statistics should continue to add more of the data from its surveys to Federal Statistical Research Data Centers and collaborate with other government agencies on the development of procedures for streamlining data access, particularly remote access for data users.

A new initiative in this area is the Institute for Research on Innovation and Science, a research center at the University of Michigan that gathers administrative data from a consortium of member academic institutions and produces secondary research data based on those records. Currently, about 30 institutions are participating in this initiative, and various associations are encouraging broader participation.

The National Science Foundation (NSF) supports a variety of centers of excellence, which are a mechanism for fostering interdisciplinary research. One or more centers of excellence focused on science and engineering statistics could fulfill a variety of functions: (1) conducting analysis on topics of interest related to the science and engineering workforce; (2) conducting methodological research, such as the small area research described in Chapter 4; (3) performing linkages to administrative data and investigating options for expanding opportunities in this area; and (4) supporting the development of the next generation of researchers—in other words, students. While NCSES has limited in-house resources for these types of activities, centers of excellence may be feasible to fund.

RECOMMENDATION 6-2: The National Center for Science and Engineering Statistics (NCSES) should consider establishing research centers of excellence. These research centers could carry out research and dissemination activities for which NCSES does not always have the funds and could facilitate the growth of research communities around the surveys.

Given the resource constraints for in-house analysis activities in particular, developing additional programs and cultivating relationships that would support analyses conducted by other researchers would be worthwhile. In addition to research centers, NCSES could consider supporting analytic grants and establishing graduate or postdoctoral fellowships, dissertation grants/awards, or an intern program. Hosting visiting scholars for brief 1- to 2-month periods also could help expand the network of data users. An important aspect of the success of these types of programs is ensuring that quick data access is facilitated and that researchers can conduct additional analyses after their participation in the program has ended should the need arise—for example, when revising a paper for publication in a journal.

An additional low-cost option would be to host researchers from other NSF units for short periods of time to conduct substantive research. NSF has a tradition of encouraging rotating assignments, and NCSES has in the past collaborated on substantive research with other NSF units. NCSES could invite staff from other units to spend brief, 2–3 month rotations in NCSES, familiarizing themselves with NCSES data and preparing reports relevant to their respective fields. NSF units that might be interested in collaborations of this type include the Division of Graduate Education within the Directorate for Education and Human Resources, the Division of Mathematical Sciences within the Directorate for Mathematical and Physical Sciences, the Directorate for Engineering, and the Office of International Science and Engineering.

NCSES also could increase its reliance on the data collection contractors to carry out analysis tasks. Outreach is needed to encourage and support research by and on underrepresented groups, especially racial/ethnic minorities and persons with disabilities.

RECOMMENDATION 6-3: To increase awareness of the usefulness of its survey data, the National Center for Science and Engineering

Statistics (NCSES) should develop programs that support data analyses conducted by other researchers. Such programs could include graduate or postdoctoral fellowships, dissertation awards, a visiting scholar program, an intern program, collaborations with researchers from other National Science Foundation divisions, and analytic grants. NCSES also could commission from its data collection contractors analysis tasks focused on dissemination goals. Special emphasis should be placed on supporting research on underrepresented groups, especially racial/ethnic minorities and persons with disabilities.

An important factor in the satisfaction of data users is thorough documentation of the data, including information on the complex sample designs, and a clear discussion of the limitations of the data. Inadequate documentation can lead to analytic errors in research that relies on secondary analysis of the microdata (for a meta-analysis of these types of errors in secondary research using the NCSES surveys, see West et al. [2016]). NCSES could consider hosting an online forum to facilitate discussion among researchers with questions related to data analysis. Existing online communities that have well-established mechanisms for sharing information about the unique characteristics of a dataset, sharing code, and discussing best practices could serve as models.[1] Online forums require an initial investment and some active engagement on the part of staff to encourage participation and answer questions, but once the communities have been established, they offer several benefits. The questions and answers are preserved in archives, so staff should not have to answer the same question repeatedly. In well-functioning online communities, moreover, users can answer each other's questions in some cases, so the staff burden is further reduced. In addition, online forums can serve as vehicles for announcements (for example, about new datasets, new reports, upcoming conferences, and so on), and also can be a source of input to NCSES about matters of interest to data users.

NCSES indicated that one of its long-term goals is to support longitudinal research with the data from its science and engineering workforce surveys, and this goal could be a focus of a center of excellence. Although some of the existing weighting challenges are unique to the NCSES surveys, the agency could benefit from input from those who have conducted other

---

[1] Examples of online support communities include the forums associated with providers of online courses, such as Udacity (https://discussions.udacity.com), and the forums for Stan, an open-source statistical analysis software package (http://discourse.mc-stan.org/categories); TensorFlo, an open-source software package for machine learning, released by Google (https://groups.google.com/a/tensorflow.org/forum/#!forum/discuss); Ubuntu, an open-source operating system (https://ubuntuforums.org); and Trellis, the platform created by the American Association for the Advancement of Science. All URLs active as of January 2018.

longitudinal studies—such as the Health and Retirement Study, the National Longitudinal Surveys, and the Wisconsin Longitudinal Study—about the types of issues they encountered and their experiences with enabling data access.

RECOMMENDATION 6-4: The National Center for Science and Engineering Statistics should continue making its work toward enabling longitudinal research based on the survey data a priority and should seek input from other agencies that provide access to longitudinal data.

RECOMMENDATION 6-5: The National Center for Science and Engineering Statistics should continue to prioritize the timeliness of its data releases and the provision of clear documentation of the data and should consider hosting an online forum or providing some other mechanism that can facilitate discussion among users of the survey data.

PUBLICATIONS

Because the *SEI* and *WMPD* reports are congressionally mandated, their publication is a priority for NCSES. The reports have been transitioned to electronic-only form in recent years, but NCSES still publishes the *Digest* versions of these reports in hard-copy form. This approach likely represents an appropriate compromise reflecting the different functions served by the full-length and condensed versions. Distributing the hard-copy *Digest* can increase the visibility of these reports, while the full electronic reports can be browsed for additional detail. Accessing the reports online also draws attention to the wealth of additional resources, including data tables, available on the NCSES Website.

The *SEI* and *WMPD* reports are updated every 2 years. With the increasing availability of data from a variety of sources at a rapid pace and the associated changes in expectations, the most important indicators likely will have to continue to be released with at least this frequency. In general with respect to survey data, whether in reports, tables, or other forms, timeliness is a need expressed repeatedly by stakeholders.

With the anticipated enhancements to the longitudinal data for the SDR, integrating longitudinal analyses into the *SEI* and *WMPD* reports could enhance the usefulness of these reports, and perhaps stimulate increased appreciation and use of the longitudinal microdata. These types of analyses conducted by the staff would also further facilitate the staff's understanding of the strengths and limitations of the longitudinal data from a user perspective.

RECOMMENDATION 6-6: The National Center for Science and Engineering Statistics should facilitate the presentation of results of longitudinal analyses in the *Science and Engineering Indicators* and *Women, Minorities, and Persons with Disabilities in Science and Engineering* reports.

In the past, NCSES used e-mail blasts to announce the release of new reports or other data products. When used correctly, maintaining a mailing list to communicate actively with stakeholders can be an efficient way to increase the visibility and use of data products. Continuing these mailings is likely worthwhile, and their impact could be enhanced by highlighting the usefulness of the data. For example, the messages could include two or three findings likely to be of particular interest to policy makers or the media.

An area in which uses of the data and communications with stakeholders could be expanded is state-level policy making. For example, there is interest in the number of students educated in science and engineering fields in public colleges and universities and the number of students who subsequently enter the workforce in the same state. NCSES already produces "state profiles," and these could be disseminated more actively to state representatives, congressional delegations, governors' offices, the National Governors Association, coordination boards, and state data centers. One workshop participant suggested that developing a smartphone app would be particularly useful for introducing busy state policy makers to key data elements because anyone could quickly pull up relevant data during a conversation with a lawmaker.

Although government employees' ability to attend professional association meetings is often limited, participation in these types of meetings can increase awareness of the NCSES surveys. In addition, many events, particularly in the Washington, DC, area, such as meetings of the American Association for the Advancement of Science, can provide opportunities to maintain staff awareness of new policy issues, the research landscape, and data needs, which can guide the prioritization of proactive dissemination efforts. Attendance at professional meetings also can provide ideas for increased communications about the data from other agencies and organizations conducting similar large-scale studies. For example, Understanding Society, the UK Household Longitudinal Study, has a well-developed outreach effort, including a policy unit specifically focused on expanding use of the data to inform policy. The unit collaborates with government units in the United Kingdom, organizations, and businesses to help them use the data effectively; organizes policy events and workshops; and conducts policy-specific training for data users. Study staff host an online user support forum and launched a podcast series that features interviews with the study team, researchers using the data, and others benefiting from the data.

As discussed, outreach efforts such as Webinars on the *SEI* and *WMPD* reports can be effective dissemination tools. Although developing the presentation materials for a Webinar can be resource-intensive, the materials once developed could be used in a variety of settings—for example, briefings for congressional staffers or meetings of professional associations, such as the American Association for the Advancement of Science.

Of particular importance is outreach to underrepresented groups. NCSES has collaborated with the National Association for Equal Opportunity in Higher Education and participated in conferences and meetings focused on issues specific to underrepresented communities. These relationships are important to ensure that the agency's surveys and reports, such as the *WMPD* reports, remain relevant to underrepresented groups. Response rates to the surveys are already low among minority-serving institutions and individual sample members (for example, historically the response rate to the NSCG has been approximately 10 percent lower among African Americans than overall), and this gap could increase without adequate attention in this area.

RECOMMENDATION 6-7: The National Center for Science and Engineering Statistics should evaluate options for expanding communications with former, existing, and new stakeholders to raise awareness of the usefulness of the survey data. State and local governments warrant special consideration, as do groups historically underrepresented in the science and engineering workforce.

References

Aanerud, R., Homer, L., Nerad, M., and Cerny, C. (2006). Paths and perceptions: Assessing doctoral education using career path analysis. In P. L. Maki and N. Borkowski (Eds.), *The Assessment of Doctoral Education: Emerging Criteria and New Models for Improving Outcomes* (pp. 109–143). Sterling, VA: Stylus.

Aanerud, R., Homer, L., Rudd, E., Morrison, E., Nerad, M., and Cerny, J. (2007). Widening the lens on gender and tenure: Looking beyond the academic labor market. *NWSA Journal, 19*(3), 105–123.

Angeles, D. (2016). Younger baby boomers and number of jobs held. *Career Outlook,* June. Available: https://www.bls.gov/careeroutlook/2016/data-on-display/younger-baby-boomers-and-number-of-jobs-held.htm [November 20, 2017].

Appelt, S., van Beuzekom, B., Galindo-Rueda, F., and de Pinho, R. (2015). *Which Factors Influence the International Mobility of Research Scientists?.* OECD Science, Technology and Industry Working Papers, 2015/02. Paris, France: OECD. doi:10.1787/5js1tmrr2233-en.

Auriol, L., Misu, M., and Freeman, R. A. (2013). *Careers of Doctorate Holders: Analysis of Labour Market and Mobility Indicators.* OECD Science, Technology and Industry Working Papers, 2013/04. Paris, France: OECD. doi:10.1787/5k43nxgs289w-en.

Biemer, P. P., and Lyberg, L. E. (2003). *Introduction to Survey Quality.* Hoboken, NJ: John Wiley & Sons.

Blanpied, W. A. (Ed.). (1995). Memoranda for program for the National Science Foundation, February 15, 1951. In W. A. Blanpied (Ed.), *Impacts of the Early Cold War on the Formulation of U.S. Science Policy: Selected Memoranda of William T. Golden, October 1950–April 1951* (pp. 90–93). Washington, DC: American Association for the Advancement of Science. Available: https://www.aaas.org/sites/default/files/reports/golden.pdf [September 20, 2017].

Blumenfield, T., and Nerad, M. (2012). International assessment: Developing a research agenda for (post)graduate education and collaboration. *Australian Universities Review, 54*(1), 72–83.

Boselli, B., and Galindo-Rueda, F. (2016). *Drivers and Implications of Scientific Open Access Publishing: Findings from a Pilot OECD International Survey of Scientific Authors.* OECD Science, Technology and Industry Policy Papers, No. 33. Paris, France: OECD. doi.org/10.1787/5jlr2z70k0bx-en.

Bush, V. (1945 [1990]). *Science—The Endless Frontier: A Report to the President on a Program for Postwar Scientific Research.* Washington, DC: National Science Foundation.

Ceci, S. J., and Williams, W. M. (2010). *The Mathematics of Sex: How Biology and Society Conspire to Limit Talented Women and Girls.* New York: Oxford University Press.

Centers for Disease Control and Prevention. (2015). *2013–2015 National Survey of Family Growth (NSFG): Sample Design Documentation.* Available: https://www.cdc.gov/nchs/data/nsfg/NSFG_2013-2015_Sample_Design_Documentation.pdf [September 20, 2017].

Chesley, K. (2015). New website reveals Ph.D. career paths for Stanford alumni. *Stanford News*, October 13.

Commission on Evidence Based Policy Making. (2017). *The Promise of Evidence-Based Policymaking.* Available: https://www.cep.gov/content/dam/cep/report/cep-final-report.pdf [November 20, 2017].

Ehrenberg, E., and Kuh, C. (Eds.). (2009). *Doctoral Education and the Faculty of the Future.* Ithaca, NY: Cornell University Press.

European Science Foundation. (2013). *Research Careers in Europe. Landscape and Horizons.* Available: http://archives.esf.org/fileadmin/Public_documents/Publications/moforum_research_careers.pdf [September 20, 2017].

European Science Foundation. (2015). *Career Tracking of Doctorate Holders.* Available: http://www.esf.org/uploads/media/Career_Tracking_01.pdf [September 20, 2017].

Fecso, R. S., Frase, M. J., and Kannankutty, N. (2012). *Using the American Community Survey as the Sampling Frame for the National Survey of College Graduates.* NCSES 12-201 Working Paper. Available: https://www.nsf.gov/statistics/ncses12201/pdf/ncses12201.pdf [September 20, 2017].

Fiegener, M., and Finamore, J. (2016). *Specific Issues for the Study.* Presentation prepared for the Panel to Evaluate the National Center for Science and Engineering Statistics Approach to Measuring the Science and Engineering Workforce, Washington, DC, April 25.

Finamore, J. (2016). *Overview of the NCSES S&E Workforce Surveys.* Presentation prepared for the Panel to Evaluate the National Center for Science and Engineering Statistics Approach to Measuring the Science and Engineering Workforce, Washington, DC, April 25.

Finamore, J., and Dillman, D. (2013). *How Mode Sequence Affects Responses by Internet, Mail, and Telephone in the National Survey of College Graduates.* Paper presented at the 5th Conference of the European Survey Research Association, Ljubljana, Slovenia.

Finn, M. (2010). *Stay Rates of Foreign Doctorate Recipients from U.S. Universities, 2007.* Oak Ridge, TN: Oak Ridge Institute for Science Education. Available: https://orise.orau.gov/stem/reports/stay-rates-foreign-doctorate-recipients-2007.pdf [September 20, 2017].

Finn, M. (2012). *Stay Rates of Foreign Doctorate Recipients from U.S. Universities, 2009.* Oak Ridge, TN: Oak Ridge Institute for Science Education. Available: https://orise.orau.gov/stem/reports/stay-rates-foreign-doctorate-recipients-2009.pdf [September 20, 2017].

Finn, M. (2014). *Stay Rates of Foreign Doctorate Recipients from U.S. Universities, 2011.* Oak Ridge, TN: Oak Ridge Institute for Science Education. Available: https://orise.orau.gov/stem/reports/stay-rates-foreign-doctorate-recipients-2011.pdf [September 20, 2017].

Freeman R., and Goroff, D. L. (2009). *Science and Engineering Careers in the United States.* Chicago, IL: University of Chicago Press.

Galarneau, D., Hinchley, C., and Ntwari, A. (2017). *Labour Market Outcomes of Graduates from Universities in the Maritime Provinces.* Available: http://www.statcan.gc.ca/pub/75-006-x/2017001/article/14798-eng.htm [September 20, 2017].

Gibbs, K. D., Jr., McGready, J., Bennett, J. C., and Griffin, K. (2014). Biomedical science Ph.D. career interest patterns by race/ethnicity and gender. *PLoS One, 9*(12), e114736.

Golde, C. M., and Dore, T. M. (2001). *At Cross Purposes: What the Experiences of Doctoral Students Reveal about Doctoral Education.* Available: http://www.phd-survey.org/report%20final.pdf [September 20, 2017].

Goulden, M., Frasch, K., and Mason, M. A. (2009). *Staying Competitive: Patching America's Leaky Pipeline in the Sciences.* Available: https://cdn.americanprogress.org/wp-content/uploads/issues/2009/11/pdf/women_and_sciences.pdf [September 20, 2017].

Gupta, D., Nerad, M., and Cerny, J. (2003). International Ph.D.s: Exploring the choice to stay or return. *International Higher Education,* Spring. Available: http://www.education.uw.edu/cirge/wp-content/uploads/2008/07/international-phds.htm [September 20, 2017].

Hastie, T., Tibshirani, R., and Friedman, J. (2009). *The Elements of Statistical Learning: Data Mining, Inference, and Prediction* (2nd Edition). New York: Springer.

Leggon, C. B., and Pearson, Jr., W. (2008). Assessing programs to improve minority participation in STEM fields: What we know and what we need to know. In R. Ehrenberg and C. Kuh (Eds.), *Doctoral Education and the Faculty of the Future* (pp. 160–174). Ithaca, NY: Cornell University Press.

Lin, Z., Pearce, R., and Wang, W. (2009). Imported talents: Demographic characteristics, achievement and job satisfaction of foreign born full time faculty in four-year American colleges. *Higher Education: The International Journal of Higher Education and Educational Planning, 57*(6), 703–721. doi:10.1007/s10734-008-9171-z.

Malcom, L., and Malcom, S. (2011). The double bind: The next generation. *Harvard Educational Review, 81*(2), 162–171.

Martin, D. E., and Pearson, Jr., W. (2005). *Preparing the STEM Workforce of the 21st Century: Broadening Participation through a Comprehensive Integrative System.* Final Workshop Report. Arlington, VA: National Science Foundation. Available: http://www2.seas.gwu.edu/~stem/report.htm [September 20, 2017].

Martinez, A., Epstein, C., and Parsad, A. (2015). *Evaluation of the National Science Foundation's Partnerships for International Research and Education (PIRE) Program.* Final Report. Cambridge, MA: Abt Associates. Available: http://abtassociates.com/AbtAssociates/files/54/541407a8-485e-4f5d-8697-bb17542544e8.pdf [September 20, 2017].

Morrison, E., Rudd, E., and Nerad, M. (2011). Onto, up, off the academic faculty ladder: The gendered effects of family on career transitions for a cohort of social science Ph.D.s. *The Review of Higher Education, 34*(4), 525–553.

Musselin, C. (2010). *The Market for Academics.* New York: Routledge.

National Academies of Sciences, Engineering, and Medicine. (2016a). *Barriers and Opportunities for 2-Year and 4-Year STEM Degrees: Systemic Change to Support Diverse Student Pathways.* Washington, DC: The National Academies Press. doi:10.17226/21739.

National Academies of Sciences, Engineering, and Medicine. (2016b). *Reducing Response Burden in the American Community Survey: Proceedings of a Workshop.* Washington, DC: The National Academies Press. doi.org/10.17226/23639.

National Academies of Sciences, Engineering, and Medicine. (2017). *Federal Statistics, Multiple Data Sources, and Privacy Protection: Next Steps.* Washington, DC: The National Academies Press. doi.org/10.17226/24893.

National Academy of Sciences, National Academy of Engineering, and Institute of Medicine. (2010). *Rising Above the Gathering Storm, Revisited: Rapidly Approaching Category 5.* Washington, DC: The National Academies Press. doi.org/10.17226/12999.

National Academy of Sciences, National Academy of Engineering, and Institute of Medicine. (2011). *Expanding Underrepresented Minority Participation: America's Science and Technology Talent at the Crossroads.* Washington, DC: The National Academies Press. doi:10.17226/12984.

National Center for Science and Engineering Statistics. (2017a). *The 2015 Survey of Doctorate Recipients Expands Its Population Coverage and Reporting on Field of Study.* National Science Foundation, InfoBrief NSF 17-319. Available: https://www.nsf.gov/statistics/2017/nsf17319 [September 20, 2017].

National Center for Science and Engineering Statistics. (2017b). *2017–2019 Survey of Graduate Students and Postdoctorates in Science and Engineering: OMB Supporting Statement Section A*. National Science Foundation. Available: https://www.reginfo.gov/public/do/PRAViewDocument?ref_nbr=201709-3145-001 [November 9, 2017].

National Center for Science and Engineering Statistics (2017c). *Supporting Statement for Survey of Earned Doctorates 2018 and 2019 Survey Cycles*. National Science Foundation. Available: https://www.reginfo.gov/public/do/PRAViewDocument?ref_nbr=201704-3145-001 [November 9, 2017].

National Center for Science and Engineering Statistics. (2017d). *Supporting Statement for the 2017 National Survey of College Graduates*. National Science Foundation. Available: https://www.reginfo.gov/public/do/PRAViewDocument?ref_nbr=201612-3145-001 [November 9, 2017].

National Center for Science and Engineering Statistics. (2017e). *Supporting Statement for the 2017 Survey of Doctorate Recipients*. National Science Foundation. Available: https://www.reginfo.gov/public/do/PRAViewDocument?ref_nbr=201704-3145-003 [October 11, 2017].

National Center for Science and Engineering Statistics. (2017f). *Survey of Doctorate Recipients: 2015 Sample Expansion and Longitudinal Analysis Capabilities*. Presentation for a meeting with representatives from the American Economic Association, Washington, DC, March 7.

National Center for Science and Engineering Statistics. (2017g). *Women, Minorities, and Persons with Disabilities in Science and Engineering: 2017*. Special Report NSF 17-310. Available: https://www.nsf.gov/statistics/2017/nsf17310/static/downloads/nsf17310-digest.pdf [December 2017].

National Research Council. (1989). *Surveying the Nation's Scientists and Engineers: A Data System for the 1990s*. Panel to Study the NSF Scientific and Technical Personnel Data System. C. F. Citro and G. Kalton (Eds.). Committee on National Statistics, Commission on Behavioral and Scoical Sciences and Education. Washington, DC: National Academy Press.

National Research Council. (2003). *Improving the Design of the Scientists and Engineers Statistical Data System*. Committee on National Statistics, Committee to Review the 2000 Decade Design of the Scientists and Engineers Statistical Data System (SESTAT), Division of Behavioral and Social Sciences and Education. Washington, DC: The National Academies Press.

National Research Council. (2008). *Using the American Community Survey for NSF's S&E Workforce Statistics Program*. Panel on Assessing the Benefits of the American Community Survey for the NSF Division of Science Resources Statistics, Committee on National Statistics, Division of Behavioral and Social Sciences and Education. Washington, DC: The National Academies Press.

National Research Council. (2012a). *Continuing Innovation in Information Technology.* Committee on Depicting Innovation in Information Technology, Computer Science and Telecommunications Board, Division on Engineering and Physical Sciences. Washington, DC: The National Academies Press. doi.org/10.17226/13427.

National Research Council. (2012b). *Research Universities and the Future of America: Ten Breakthrough Actions Vital to Our Nation's Prosperity and Security: Summary*. Policy and Global Affairs, Board on Higher Education and Workforce, Committee on Research Universities. Washington, DC: The National Academies Press. doi.org/10.17226/13299.

National Research Council. (2014). *Capturing Change in Science, Technology, and Innovation: Improving Indicators to Inform Policy.* Panel on Developing Science, Technology, and Innovation Indicators for the Future. R. E. Litan, A. W. Wyckoff, and K. Husbands Fealing (Eds.). Committee on National Statistics, Division of Behavioral and Social Sciences and Education; Board on Science, Technology, and Economic Policy, Division of Policy and Global Affairs. Washington, DC: The National Academies Press.

National Science Board. (2000). *Science and Engineering Indicators 2000.* NSB 00-1. Arlington, VA: National Science Foundation.

National Science Board. (2015). *Revisiting the STEM Workforce: A Companion to Science and Engineering Indicators 2014.* NSB 2015-10. Arlington, VA: National Science Foundation.

National Science Board. (2016). *Science and Engineering Indicators 2016.* NSB-2016-1. Arlington, VA: National Science Foundation.

National Science Board. (2017). *SHE Doctorates in the Workforce, 1993–2013.* Available: https://www.nsf.gov/nsb/sei/infographic2/#about [November 22, 2017].

Nerad, M. (2009). Confronting common assumptions: Designing future-oriented doctoral education. In R. Ehrenberg and C. Kuh (Eds.), *Doctoral Education and the Faculty of the Future* (pp. 80–89). Ithaca, NY: Cornell University Press.

Nerad, M., and Blumenfield, T. (2012). *Investigating the International Experiences in STEM Graduate Education and Beyond: Report from the February 2011 Workshop to Develop a Research Agenda.* Seattle, WA: University of Washington, Center for Innovation and Research in Graduate Education.

Nerad, M., and Cerny, J. (1999). Postdoctoral patterns, career advancement, and problems, in American Association for the Advancement of Science. *Science, 285*(5433), 1533–1535.

Nerad, M., and Cerny, J. (2002). Postdoctoral appointments and employment patterns of science and engineering doctoral recipients ten-plus years after Ph.D. completion: Selected results from the "Ph.D.s—Ten Years Later" Study. *Communicator, 35*(7), August–September, Available: http://www.education.uw.edu/cirge/wp-content/uploads/2008/07/postdoctoral-appointments.pdf [September 20, 2017].

Ong, M., Wright, C., Espinosa, L. L., and Orfield, G. (2011). Inside the double bind: A synthesis of empirical research on undergraduate and graduate women of color in science, technology, engineering, and mathematics. *Harvard Educational Review, 81*(2), 172–209.

Pearson, Jr., W., Frehill, L., and McNeely, C. (Eds.). (2015). *Advancing Women in Science: An International Perspective.* New York: Springer.

Polka, J. (2014). *Where Will a Biology Ph.D. Take You?* Infographic, American Society for Cell Biology. Available: http://www.ascb.org/compass/compass-points/where-will-a-biology-phd-take-you [September 19, 2017].

Pressman, L., Planting, M., Yuskavage, R. Okubo, S., Moyland, C. and Bond, J. (2017). *The Economic Contribution of University/Nonprofit Inventions in the United States: 1996–2015.* Available: https://www.autm.net/AUTMMain/media/Partner-Events/Documents/Economic-_Contribution_University-Nonprofit_Inventions_US_1996-2015_BIO_AUTM.pdf [September 20, 2017].

Rao, J. N. K., and Molina, I. (2015). *Small Area Estimation.* New York: John Wiley & Sons.

Redline, C. (2013). Clarifying survey questions in a web survey. *Public Opinion Quarterly, 77*(Special Issue), 89–105.

Rivers, E. (2016). *The Current NCSES Approach and an Overview of the NCSES Science and Engineering Education Surveys.* Presentation prepared for the Panel to Evaluate the National Center for Science and Engineering Statistics Approach to Measuring the Science and Engineering Workforce, Washington, DC, April 25.

Roessner, D., Bond, J., Okubo, S., and Planting, M. (2017). The economic impact of licensed commercialized inventions originating in university research. *Research Policy, 42*(1), 23–34.

Rubio, D. M., Primack, B. A., Switzer, G. E., Bryce, C. L., Seltzer, D. L., and Kapoor, W. N. (2011). A comprehensive career-success model for physician-scientists. *Academic Medicine, 86*(12), 1571–1576. doi:10.1097/ACM.0b013e31823592fd.

Rudd, E., Nerad, M., Emory, M., and Picciano, J. (2008). *CIRGE Spotlight #2 on Doctoral Education: Professional Development for Ph.D. Student: Do They Really Need It? Findings from Social Science Ph.D.s—Five+ Years Out.* Seattle, WA: CIRGE. Available: http://www.education.uw.edu/cirge/wp-content/uploads/2008/10/professional-skills-spotlight_10_07_08.pdf [October 15, 2017].

Rudd, E., Picciano, J., Nerad, M., and Cerny, J. (2010). *The Influence of Postdoctoral Training on Prestige and Time in Academic Careers: Findings from Ph.D.'s—Ten Years Later, A National Survey of Ph.D. Careers.* Report to the Alfred Sloan Foundation.

Sakshaug, J. W., Wolter, S., and Kreuter F. (2015). Obtaining record linkage consent: Results from a wording experiment in Germany. *Survey Insights: Methods from the Field,* November 13. Available: http://surveyinsights.org/?p=7288 [December 2017].

Sakshaug, J. W., Hülle, S., Schmucker, A., and Liebig, S. (2017). Exploring the effects of interviewer- and self-administered survey modes on record linkage consent rates and bias. *Survey Research Methods, 11*(2). doi:10.18148/srm/2017.v11i2.7158.

Schuman, H. (1966). The random probe: A technique for evaluating the validity of closed questions. *American Sociological Review, 31*(2), 218–222.

Slaughter, J. B., Tao, Y., and Pearson, W. Jr. (Eds.). (2015). *Changing the Face of Engineering: The African American Experience.* Baltimore, MD: Johns Hopkins University Press.

Smith, P., Pont, M., and Jones, T. (2003). Developments in business survey methodology in the Office for National Statistics, 1994–2000. *Journal of the Royal Statistical Society: Series D (The Statistician), 52*(3), 257–286. doi:10.1111/1467-9884.03571.

Smyth, J. D., Dillman, D. A., Christian, L. M., and Stern, M. J. (2006). Comparing check-all and forced-choice question formats in web surveys. *Public Opinion Quarterly, 70*(1), 66–77.

Smyth, J. D., Christian, L. M., and Dillman, D. A. (2008). Does "yes or no" on the telephone mean the same as check-all-that-apply on the web? *Public Opinion Quarterly, 72*(1), 103–111.

Steelman, J. R. (1949). *Science and Public Policy: A Program for the Nation.* Washington, DC: U.S. Government Printing Office.

Stephan, P. E. (2009). Tracking the placement of students as a measure of technology transfer. In G. D. Libecap (Ed.), *Measuring the Social Value of Innovation: A Link in the University Technology Transfer and Entrepreneurship Equation* (pp. 113–140). Bingley, UK: Emerald Group.

Stephan, P. E. (2011). The biomedical workforce in the U.S.: An example of positive feedbacks. In C. Antonelli (Ed.), *Handbook on the Economic Complexity of Technological Change* (pp. 240–261). Cheltenham, UK: Edward Elgar.

U.S. Census Bureau. (2006). *Current Population Survey Design and Methodology.* Technical Paper 66. Available: https://www.census.gov/prod/2006pubs/tp-66.pdf [September 20, 2017].

Weisberg, A., and Galinsky, E. (2014). *Changing Families, Changing Work: A Paper for the Commemoration of the 50th Anniversary of President Kennedy's Commission on the Status of Women Report American Women.* Washington, DC: Families and Work Institute. Available: http://digitalcommons.ilr.cornell.edu/cgi/viewcontent.cgi?article=2642&context=key_workplace [December 2017].

West, B. T., Ghimire, D., and Axinn, W. G. (2015). Evaluating a modular design approach to collecting survey data using text messages. *Survey Research Methods, 9*(2), 111–123.

West, B. T., Sakshaug, J. W., and Aurelien, G. A. S. (2016). How big of a problem is analytic error in secondary analyses of survey data? *PLoS One, 11*(6), e0158120. doi:10.1371/journal.pone.0158120.

Appendix A

Figures and Tables in the 2016 *Science and Engineering Indicators* Report That Use Data from the Survey of Graduate Students and Postdoctorates in Science and Engineering, the Survey of Earned Doctorates, the National Survey of College Graduates, and the Survey of Doctorate Recipients

2016 *Science and Engineering Indicators* Figure or Table	Title of the Figure or Table	NCSES Survey Providing the Data	NCSES Measures Providing the Data
Table 2-3	U.S. citizen and permanent resident S&E doctorate recipients whose baccalaureate origin is a high Hispanic enrollment institution, by ethnicity and race: 2009–2013	SED	Doctorate recipients' primary source of support
Table 2-4	U.S. citizen and permanent resident S&E doctorate recipients whose baccalaureate origin is an HBCU, by ethnicity and race: 2009–2013	SED	U.S. S&E doctorate recipients who are U.S. citizens or permanent residents and whose baccalaureate is from an HBCU
Table 2-5	U.S. citizen and permanent resident S&E doctorate recipients who reported earning college credit from a community or 2-year college, by ethnicity and race: 2009–2013	SED	S&E doctorate recipients who are U.S. citizens or permanent residents and report earning college credit from a community or 2-year college
Table 2-6	Community college attendance among recent recipients of S&E degrees, by sex, race, ethnicity, citizenship status, and parents' education level: 2013	NSCG	Community college attendance among recent recipients of S&E degrees by sex, race, ethnicity, citizenship status, and parents' education level
Figure 2-6	Full-time S&E graduate students, by field and mechanism of primary support: 2013	GSS	Full-time S&E graduate students by field and mechanism of primary support
App. Table 2-7	Full-time S&E graduate students, by source and mechanism of primary support: 1998–2013	GSS	Full-time S&E graduates by source and mechanism of primary support
Figure 2-7	Full-time S&E graduate students with primary support from federal government, by field: 2013	GSS	Full-time S&E graduates supported by the federal government by field
App. Table 2-8	Full-time S&E graduate students, by field and mechanism of primary support: 2013	GSS	Full-time S&E graduates by field and mechanism of primary support

Figure 2-8	Full-time S&E graduate students, by source of primary support: 1998–2013	GSS	Full-time S&E graduate students by source of primary support
App. Table 2-9	Full-time S&E graduate students primarily supported by federal government, by field and mechanism of primary support: 2013	GSS	Full-time S&E graduate students by field and mechanism of primary support
Table 2-10	Primary support mechanisms for S&E doctorate recipients, by 2010 Carnegie classification of doctorate-granting institution: 2013	SED	Doctorate recipients' primary source of support
App. Table 2-10	Full-time S&E graduate students primarily supported by federal government, by field: 1998–2013	GSS	Full-time S&E graduates supported by the federal government by field
App. Table 2-11	Full-time S&E graduate students primarily supported by federal government, by agency: 1998–2013	GSS	Full-time S&E graduates supported by the federal government by agency
Table 2-12	Median number of years from entering graduate school to receipt of S&E doctorate, by 2010 Carnegie classification of doctorate-granting institution: 1998–2013	SED	Median number of years from S&E doctorate recipients' entry to graduate school to receipt of doctorate by field
App. Table 2-12	Full-time S&E graduate students primarily supported by federal government, by field and agency: 2013	GSS	Full-time S&E graduates supported by the federal government by agency Full-time S&E graduates supported by the federal government by field
Table 2-13	Recipients of U.S. S&E doctorates on temporary visas, by country/economy of origin: 1993–2013	SED	U.S. S&E doctorate recipients on temporary visas
App. Table 2-13	Doctorate recipients' source of primary support, by broad field of study, sex, citizenship, race, and ethnicity: 2011–2013	SED	Doctorate recipients' primary source of support

continued

2016 *Science and Engineering Indicators* Figure or Table	Title of the Figure or Table	NCSES Survey Providing the Data	NCSES Measures Providing the Data
Table 2-14	Asian recipients of U.S. S&E doctorates on temporary visas, by field and country/economy of origin: 1993–2013	SED	Asian recipients of U.S. S&E doctorates on temporary visas
App. Table 2-14	Amount of undergraduate and graduate debt of S&E doctorate recipients, by field: 2013	SED	S&E doctorate recipients' undergraduate and graduate debt
Table 2-15	European recipients of U.S. S&E doctorates on temporary visas, by field and region/country of origin: 1993–2013	SED	European recipients of U.S. S&E doctorates on temporary visas
Table 2-16	North American, South American, and Middle Eastern recipients of U.S. S&E doctorates, by field and region/country of origin: 1993–2013	SED	North American, South American, and Middle Eastern recipients of U.S. S&E doctorates
App. Table 2-24	S&E graduate enrollment, by sex and field: 2000–2013	GSS	S&E graduate enrollment by sex and field
Figure 2-24	U.S. S&E doctoral degree recipients, by selected Asian country/economy of origin: 1993–2013	SED	Asian recipients of U.S. S&E doctorates on temporary visas
App. Table 2-25	First-time full-time S&E graduate students, by citizenship and field: 2000–2013	GSS	Full-time S&E graduates supported by the federal government by citizenship
			Full-time S&E graduates supported by the federal government by field
Figure 2-25	U.S. S&E doctoral degree recipients, by selected Western European country: 1993–2013	SED	U.S. S&E doctorate recipients by selected Western European country
App. Table 2-26	S&E graduate enrollment, by citizenship, field, race, and ethnicity: 2000–2013	GSS	S&E graduate enrollment by citizenship, field, and race or ethnicity

Figure 2-26	U.S. S&E doctoral degree recipients from Europe, by region: 1993–2013	SED	U.S. S&E doctorate recipients from Europe
Figure 2-27	U.S. S&E doctoral degree recipients from Canada, Mexico, and Brazil: 1993–2013	SED	U.S. S&E doctorate recipients from Canada, Mexico, and Brazil
App. Table 2-31	Median number of years from S&E doctorate recipients' entry to graduate school to receipt of doctorate, by field: 1983–2013	SED	Median number of years from S&E doctorate recipients' entry to graduate school to receipt of doctorate by field
App. Table 2-32	Median number of years from S&E doctorate recipients' entry to graduate school to receipt of doctorate, by sex, citizenship, race, ethnicity, and broad field of study: 2013	SED	Median number of years from S&E doctorate recipients' entry to graduate school to receipt of doctorate by sex, citizenship, race, ethnicity, and broad field of study
Figure 3-1	Employment in S&E occupations, by broad occupational category: 2013 and 2014	SESTAT	Employment in S&E occupations by broad occupational category SESTAT
Figure 3-2	S&E degrees among college graduates, by field and level of highest degree: 2013	SESTAT	S&E degrees among college graduates by field and level of highest degree
Figure 3-3	Measures and size of the U.S. S&E workforce: 2013 and 2014	NSCG	Job requires S&E technical expertise at bachelor's level
			Employed in S&E occupations (SESTAT)
		SESTAT	Employed workers with at least one degree in S&E field
			Job relatedness to highest degree for those with highest degree in S&E field by occupation
App. Table 3-3	Scientists and engineers, by occupation and degree field: 2013	SESTAT	Scientists and engineers by occupation and field of highest degree
Table 3-4	Educational background of college graduates employed in S&E occupations, by broad S&E occupational category: 2013	SESTAT	Educational background of college graduates employed in S&E occupations by broad S&E occupational category

continued

2016 *Science and Engineering Indicators* Figure or Table	Title of the Figure or Table	NCSES Survey Providing the Data	NCSES Measures Providing the Data
App. Table 3-4	Employment sector of S&E highest degree holders and workers in S&E occupations: 1993, 2003, and 2013	SESTAT	Employment sector of S&E highest degree holders Employment sector of workers in S&E occupations
Figure 3-4	Average annual growth in the total number of employed individuals with highest degree in S&E, by field and level of highest degree: 2003–2013	SESTAT	Employed individuals with highest degree in S&E by field and level of highest degree: 2003 Employed individuals with highest degree in S&E by field and level of highest degree: 2013
Table 3-5	Relationship of highest degree to job among S&E highest degree holders not in S&E occupations, by degree level: 2013	SESTAT	Relationship of highest degree to job among S&E highest degree holders not in S&E occupations
App. Table 3-5	Employment sector of S&E highest degree holders, by level and field of highest degree: 2013	SESTAT	Employment sector of S&E highest degree holders by level and field of highest degree
Table 3-6	Employment sector of scientists and engineers, by broad occupational category and degree field: 2013	SESTAT	Employment sector of scientists and engineers by broad occupational category and degree field
App. Table 3-6	Scientists and engineers participating in work-related training, by labor force status, highest degree level, and sex: 2013	SESTAT	Scientists and engineers participating in work-related training by labor force status, highest degree level, and sex
Figure 3-6	Occupational distribution of scientists and engineers, by broad field of highest degree: 2013	SESTAT	Occupational distribution of scientists and engineers by broad field of highest degree
Table 3-7	Self-employed scientists and engineers, by education, occupation, and type of business: 2013	SESTAT	Self-employed scientists and engineers by education, occupation, and type of business
App. Table 3-7	Most important reason for scientists and engineers to participate in work-related training, by labor force status: 2013	SESTAT	Most important reason for scientists and engineers to participate in work-related training by labor force status

Figure 3-7	Occupational distribution of S&E highest degree holders, by field of highest degree: 2013	SESTAT	Occupational distribution of scientists and engineers by broad field of highest degree
App. Table 3-8	Unemployment rates of scientists and engineers, by level of highest degree and broad occupational category: 2003–2013	SESTAT	Unemployment rates of scientists and engineers by level of highest degree and broad occupational category
Figure 3-8	S&E degree holders working in S&E occupations, by level and field of S&E highest degree: 2013	SESTAT	Percent of S&E degree holders working in S&E occupations by level and field of S&E highest degree
App. Table 3-9	Median annual salaries for employed college-educated individuals, by broad field and level of highest degree and broad occupational category: 2013	NSCG	Median annual salaries for employed college-educated individuals in S&E occupations by broad field and level of highest degree: all degrees, S&E-related degrees, and non-S&E degrees
			Median annual salaries for employed college-educated individuals in S&E occupations with highest degrees in S&E by level of highest degree
			Median annual salaries for employed college-educated individuals with highest degrees in S&E by level of highest degree and broad occupational category: all occupations, S&E-related occupations, and non-S&E occupations
			Median annual salaries for employed college-educated individuals by broad field and level of highest degree and broad occupational category: all fields, S&E-related fields, and non-S&E fields for all occupations, S&E-related occupations, and non-S&E occupations
Figure 3-9	S&E degree holders employed in jobs related to highest degree, by level of and years since highest degree: 2013	SESTAT	S&E degree holders employed in jobs related to highest degree by level of and years since highest degree

continued

2016 *Science and Engineering Indicators* Figure or Table	Title of the Figure or Table	NCSES Survey Providing the Data	NCSES Measures Providing the Data
Table 3-10	R&D activity rate of scientists and engineers employed in S&E occupations, by broad occupational category and level of highest degree: 2013	SESTAT	Scientists and engineers employed in S&E occupations and active in R&D by broad occupational category and level of highest degree Scientists and engineers employed in S&E occupations by broad occupational category and level of highest degree
App. Table 3-10	Postgraduation plans of doctorate recipients with definite commitments, by broad field of study: Selected years, 1973–2013	SED	Postgraduation plans of doctorate recipients with definite commitments by broad field of study: 1973, 1983 Postgraduation plans of doctorate recipients with definite commitments by broad field of study: 1993–2013
Figure 3-10	S&E highest degree holders, by degree level and employment sector: 2013	SESTAT	S&E highest degree holders by degree level and employment sector
Table 3-11	Scientists and engineers participating in work-related training, by labor force status and occupation: 2013	SESTAT	Scientists and engineers participating in work-related training by labor force status and occupation
App. Table 3-11	Age profile of employed scientists and engineers, by sex and field of highest degree: 2013	SESTAT	Age profile of employed scientists and engineers by sex and field of highest degree
Figure 3-11	Broad S&E occupational categories, by employment sector: 2013	SESTAT	Broad S&E occupational categories by employment sector

Table 3-12	Scientists and engineers who are working involuntarily out of field, by S&E degree field: 2003–2013	SESTAT	Scientists and engineers who are working involuntarily out of field by S&E degree field
App. Table 3-12	Employed scientists and engineers, by sex and occupation: 2013	SESTAT	Employed scientists and engineers by sex and occupation
Figure 3-12	Scientists and engineers employed in the business sector, by employer size: 2013	SESTAT	Scientists and engineers employed in the business sector by employer size
App. Table 3-13	Employed S&E highest degree holders, by sex and field of degree: 2013	SESTAT	Employed S&E highest degree holders by sex and field of degree
Figure 3-13	Employed scientists and engineers with R&D activity, by broad field of highest degree and broad occupational category: 2013	SESTAT	Employed scientists and engineers by broad field of highest degree and broad occupational category Employed scientists and engineers with R&D activity by broad field of highest deree and broad occupational category
Table 3-14	Labor market indicators for recent S&E degree recipients up to 5 years after receiving degree, by level and field of highest degree: 2013	SESTAT	Involuntary out-of-field rate for recent S&E degree recipients up to 5 years after receiving degree by level and field of highest degree Median annual salary for recent S&E degree recipients up to 5 years after receiving degree by level and field of highest degree Unemployment rate for recent S&E degree recipients up to 5 years after receiving degree by level and field of highest degree

continued

2016 *Science and Engineering Indicators* Figure or Table	Title of the Figure or Table	NCSES Survey Providing the Data	NCSES Measures Providing the Data
Table 3-14	Employed S&E highest degree holders, by sex, race, ethnicity, field of highest degree, and broad occupational category: 2013	SESTAT	Employed S&E highest degree holders by race, field of highest degree, and broad occupational category Employed S&E highest degree holders by sex, field of highest degree, and broad occupational category
Figure 3-14	Employed SEH doctorate holders with R&D activity, by years since doctoral degree: 2013	SDR	Employed SEH doctorate holders with R&D activity by years since doctoral degree
Table 3-15	Employment characteristics of recent SEH doctorate recipients up to 3 years after receiving doctorate, by field of degree: 2001–2013	SDR	Employment characteristics of recent SEH doctorate recipients up to 3 years after receiving doctorate by field of degree: 2001–2013
App. Table 3-15	Employed scientists and engineers, by race, ethnicity, and occupation: 2013	SESTAT	Employed scientists and engineers by race, ethnicity, and occupation
Figure 3-15	Unemployment rates of S&E highest degree holders, by level of and years since highest degree: 2013	SESTAT	Unemployment rates of S&E highest degree holders by level of and years since highest degree
Table 3-16	Employed SEH doctorate recipients holding tenure and tenure-track appointments at academic institutions, by field of and years since degree: 1993–2013	SDR	Employed SEH doctorate recipients holding tenure and tenure-track appointments at academic institutions by field of and years since degree: 1993–2013
App. Table 3-16	Employed S&E highest degree holders, by race, ethnicity, and field of degree: 2013	SESTAT	Employed S&E highest degree holders by race, ethnicity, and field of degree
Table 3-17	Median salaries for recent SEH doctorate recipients up to 5 years after receiving degree, by field of degree and employment sector: 2013	SDR	Median salaries for recent SEH doctorate recipients up to 5 years after receiving degree by field of degree and employment sector

App. Table 3-17	Estimate and median salary of full-time workers with highest degree in S&E field, by sex and occupation: 2013	SESTAT	Estimate of full-time workers with highest degree in S&E field by sex and occupation
			Median salary of full-time workers with highest degree in S&E field by sex and occupation
Table 3-18	Median salaries for recent U.S. SEH doctorate recipients in postdoc and non-postdoc positions up to 5 years after receiving degree: 2013	SDR	Median salaries for recent U.S. SEH doctorate recipients in postdoc and nonpostdoc positions up to 5 years after receiving degree
App. Table 3-18	Estimate and median salary of full-time workers with highest degree in S&E field, by race, ethnicity, and occupation: 2013	SESTAT	Estimate of full-time workers with highest degree in S&E field by race, ethnicity, and occupation
			Median salary of full-time workers with highest degree in S&E field by race, ethnicity, and occupation
Figure 3-18	S&E highest degree holders who are working involuntarily out of field, by field of and years since highest degree: 2013	SESTAT	S&E highest degree holders who are working involuntarily out of field by field of and years since highest degree
Table 3-19	Racial and ethnic distribution of U.S. residents, and of employed individuals in S&E occupations, with S&E degrees, and with college degrees: 2013	NSCG	Racial and ethnic distribution of employed individuals with college degrees
		SESTAT	Racial and ethnic distribution of employed individuals in S&E occupations
			Racial and ethnic distribution of employed individuals with highest degree in S&E
App. Table 3-19	Race and ethnic distribution of workers in S&E occupations, by nativity: 2013	SESTAT	Race and ethnic distribution of workers in S&E occupations by nativity
Figure 3-19	Median salaries for employed college-educated individuals, by broad field of highest degree and years since highest degree: 2013	SESTAT	Median salaries for employed college-educated individuals by broad field of highest degree and years since highest degree

continued

2016 *Science and Engineering Indicators* Figure or Table	Title of the Figure or Table	NCSES Survey Providing the Data	NCSES Measures Providing the Data
Table 3-20	Distribution of workers in S&E occupations, by race and ethnicity: 1993–2013	SESTAT	Distribution of workers in S&E occupations by race and ethnicity
Figure 3-20	Median salaries for S&E highest degree holders, by broad field of and years since highest degree: 2013	SESTAT	Median salaries for S&E highest degree holders by broad field of and years since highest degree
Table 3-21	Racial and ethnic distribution of employed individuals with S&E highest degree, by field of highest degree: 2013	SESTAT	Racial and ethnic distribution of employed individuals with highest degree in S&E
App. Table 3-21	Plans of foreign recipients of U.S. doctorates to stay in the United States, by field of doctorate and place of origin: 2002–2013	SED	Plans of foreign recipients of U.S. doctorates to stay in the United States by field of doctorate and place of origin
Figure 3-21	Median salaries for S&E highest degree holders, by level of and years since highest degree: 2013	SESTAT	Median salaries for S&E highest degree holders by level of and years since highest degree
Table 3-22	Racial and ethnic distribution of employed individuals with S&E highest degree, by level of highest degree: 2013	SESTAT	Racial and ethnic distribution of employed individuals with S&E highest degree by level of highest degree: 2013
Figure 3-22	Recent U.S. SEH doctorate recipients in postdoc positions, by field of and years since doctorate: 2013	SESTAT	Recent U.S. SEH doctorate recipients in postdoc positions by field of and years since doctorate
Table 3-23	Racial and ethnic distribution of employed women in S&E occupations and with S&E highest degrees: 1993 and 2013	SESTAT	Racial and ethnic distribution of employed women in S&E occupations Racial and ethnic distribution of employed women with S&E highest degrees

Figure 3-23	Age distribution of scientists and engineers in the labor force, by sex: 1993 and 2013	SESTAT	Age distribution of scientists and engineers in the labor force by sex
Table 3-24	Median annual salary among S&E highest degree holders working full time, by sex, race, and ethnicity: 1995, 2003, and 2013	SESTAT	Median annual salary among S&E highest degree holders working full-time by sex, race, and ethnicity
Figure 3-24	Age distribution of employed scientists and engineers, by broad occupational category and broad field of highest degree: 2013	SESTAT	Age distribution of employed scientists and engineers by broad occupational category and broad field of highest degree
Table 3-25	Foreign-born workers in S&E occupations, by education level: 1993, 2003, and 2013	SESTAT	Foreign-born workers in S&E occupations by education level (SESTAT)
Figure 3-26	Older scientists and engineers who report not working because of retirement, by age: 2013	SESTAT	Older scientists and engineers who report not working because of retirement by age
Figure 3-27	Women in the workforce and in S&E: 1993 and 2013	NSCG	Women as a percentage of the college-educated workforce
		SESTAT	Women as a percentage of S&E highest degree holders
			Women in S&E occupations
Figure 3-28	Women in S&E occupations: 1993–2013	SESTAT	Women in S&E occupations
Figure 3-29	Employed women with highest degree in S&E, by degree level: 1993–2013	SESTAT	Employed women with highest degree in S&E by degree level
Figure 3-30	Highest degree holders in S&E not in the labor force, by sex and age: 2013	SESTAT	Highest degree holders in S&E not in the labor force by sex and age
Figure 3-31	Estimated salary differences between women and men with highest degree in S&E employed full time, controlling for selected characteristics, by degree level: 2013	SESTAT	Estimated salary differences between women and men with highest degree in S&E employed full time, controlling for selected characteristics by degree level

continued

2016 *Science and Engineering Indicators* Figure or Table	Title of the Figure or Table	NCSES Survey Providing the Data	NCSES Measures Providing the Data
Figure 3-32	Estimated salary differences between minorities and whites and Asians with highest degree in S&E employed full time, controlling for selected characteristics, by degree level: 2013	SESTAT	Estimated salary differences between minorities and whites and Asians with highest degree in S&E employed full-time, controlling for selected characteristics by degree level
Figure 3-33	Foreign-born scientists and engineers employed in S&E occupations, by highest degree level and broad S&E occupational category: 2013	SESTAT	Foreign-born scientists and engineers employed in S&E occupations by highest degree level and broad S&E occupational category
Figure 3-34	Foreign-born individuals with highest degree in S&E living in the United States, by place of birth: 2013	SESTAT	Foreign-born individuals with highest degree in S&E living in the United States by place of birth
Figure 3-36	Plans at graduation of foreign recipients of U.S. S&E doctoral degrees to stay in the United States, by year of doctorate: 1993–2013	SED	Plans at graduation of foreign recipients of U.S. S&E doctorates to stay in the United States by year of doctorate
Table 5-12	Tenure status, by field of doctorate: 1997 and 2013	SDR	Tenure status by field of doctorate
Table 5-13	Tenured S&E doctorate holders employed in academia, by age: 1997 and 2013	SDR	Tenure status of academically employed SEH doctorate holders by age
App. Table 5-13	S&E doctorate holders employed in academia, by type of position and degree field: 1973–2013	SDR	SEH doctorate holders employed in academia by type of position and degree field
Table 5-14	Women as a percentage of S&E doctorate holders employed in academia, by position: Selected years, 1973–2013	SDR	Women as a percentage of SEH doctorate holders employed in academia by position
App. Table 5-14	S&E doctorate holders employed in academia, by type of position, sex, and degree field: 1973–2013	SDR	SEH doctorate holders employed in academia by type of position, sex, and degree field

Table 5-15	Tenured S&E doctorate holders employed in academia, by sex and field: 1997 and 2013	SDR	Tenure status of SEH doctorate holders employed in academia by sex and field
App. Table 5-15	S&E doctorate holders employed in academia, by type of position, race, ethnicity, and degree field: 1973–2013	SDR	SEH doctorate holders employed in academia by type of position, degree field, race, and ethnicity
Table 5-16	Underrepresented minorities as a percentage of S&E doctorate holders employed in academia, by position: Selected years, 1973–2013	SDR	Underrepresented minorities as a percentage of SEH doctorate holders employed in academia by position
App. Table 5-16	S&E doctorate holders employed in academia, by type of position, degree field, and citizenship: 2013	SDR	SEH doctorate holders employed in academia by type of position, degree field, and citizenship
Figure 5-16	S&E doctorate holders employed in academia, by type of position: 1973–2013	SDR	SEH doctorate holders employed in academia by position type
Table 5-17	Academically employed S&E doctorate holders, by age: 1995 and 2013	SDR	Academically employed SEH doctorate holders by age
App. Table 5-17	Age distribution of S&E doctorate holders in full-time faculty positions at research universities and other academic institutions: 1973–2013	SDR	Age distribution of SEH doctorate holders in full-time faculty positions at research universities and other academic institutions
Figure 5-17	S&E doctorate holders employed in academia, by S&E field: Selected years, 1973–2013	SDR	SEH doctorate holders employed in academia by S&E field
Table 5-18	Full-time S&E faculty reporting research as primary work activity, by years since doctorate and degree field: 2013	SDR	SEH faculty reporting research as primary work activity by years since doctorate and degree field
App. Table 5-18	S&E doctorate holders employed in academia, by research priority, type of position, and degree field: 1973–2013	SDR	SEH doctorate holders employed in academia by research priority, type of position, and degree field

continued

2016 *Science and Engineering Indicators* Figure or Table	Title of the Figure or Table	NCSES Survey Providing the Data	NCSES Measures Providing the Data
Figure 5-18	Women as a percentage of S&E doctorate holders employed full time in academia, by academic rank: Selected years, 1973–2013	SDR	Women as a percentage of SEH doctorate holders employed full time in academia by academic rank
Table 5-19	Full-time S&E graduate students and graduate research assistants at universities and colleges, by degree field: Selected years, 1973–2013	GSS	Full-time SEH graduate students and graduate research assistants at universities and colleges by degree field
App. Table 5-19	Early career S&E doctorate holders employed in academia, by Carnegie institution type, years since doctorate, and type of position: 1997–2013	SDR	Early career SEH doctorate holders employed in academia by Carnegie institution type, years since doctorate, and type of position
Figure 5-19	S&E doctorate holders employed in academia, by birthplace: 1973–2013	SDR	SEH doctorate holders employed in academia by birthplace
Table 5-20	S&E doctorate holders with academic employment in postdoc positions, by demographic group: Selected years, 1973–2013	SDR	SEH doctorate holders with academic employment in postdoc positions by demographic group
App. Table 5-20	Academic S&E doctorate holders with federal support, by degree field, research activity, and type of position: 1973–2013	SDR	Academic SEH doctorate holders with federal support by degree field, research activity, and type of position
Figure 5-20	Full-time faculty ages 65–75 at research universities and other higher education institutions: 1973–2013	SDR	Full-time faculty aged 65–75 at research universities and other higher education institutions
Table 5-21	S&E doctorate holders with academic employment in postdoc positions, by Carnegie classification of employer and years since doctorate: 2013	SDR	SEH doctorate holders with academic employment in postdoc positions by Carnegie institution type and years since doctorate

App. Table 5-21	S&E doctorate holders and full-time faculty with federal support, by degree field and Carnegie classification of employer: 2013	SDR	SEH doctorate holders and full-time faculty with federal support by degree field and Carnegie classification of employer
Figure 5-21	Primary work activity of full-time doctoral S&E faculty: Selected years, 1973–2013	SDR	Job duties of full-time doctoral SEH faculty
App. Table 5-22	S&E doctorate holders with federal support, by Carnegie class, race, ethnicity, and sex: 2013	SDR	SEH doctorate holders with federal support by Carnegie class, race and ethnicity, and gender
Figure 5-22	S&E doctorate holders with academic employment in a postdoc position, by degree field: Selected years, 1973–2013	SDR	SEH doctorate holders employed in academic postdoc positions by degree field
App. Table 5-23	Early career S&E doctorate holders employed in academia with federal support, by degree field, years since doctorate, and type of position: 1973–2013	SDR	Early career SEH doctorate holders employed in academia with federal support by degree field, years since doctorate, and type of position
Figure 5-23	S&E doctorate holders in very high research activity institutions with federal support, by sex, race, and ethnicity: 2013	SDR	Full-time SEH doctoral faculty in very high research activity institutions with federal support by gender, race, and ethnicity
Table 5-B	Foreign-trained S&E doctorate holders employed in academia, by degree field and sex: 2013	NSCG	Foreign-trained SEH doctorate holders employed in academia by degree field and sex
Table 5-C	Foreign-trained S&E doctorate holders employed in academia, by research and teaching focus: 2013	NSCG	Foreign-trained SEH doctorate holders employed in academia by research and teaching focus
Table 8-32	Employed science, engineering, and health doctorate holders as a percentage of the workforce, by state: 1993–2013	SDR	Employed SEH doctorate holders
Table 8-45	Science and engineering doctorates conferred per 1,000 employed science, engineering, and health doctorate holders, by state: 1990–2013	SDR SED	Employed SEH doctorate holders S&E doctorates conferred

continued

2016 *Science and Engineering Indicators* Figure or Table	Title of the Figure or Table	NCSES Survey Providing the Data	NCSES Measures Providing the Data
Table 8-46	Academic S&E article output per 1,000 SEH doctorate holders in academia, by state: 1999–2013	SDR	SEH doctorate holders in academia
Table 8-48	Academic patents awarded per 1,000 SEH doctorate holders in academia, by state: 1996–2014	SDR	SEH doctorate holders in academia

NOTE: GSS = Survey of Graduate Students and Postdoctorates in Science and Engineering; HBCU = historically black college or university; NSCG = National Survey of College Graduates; R&D = research and development; S&E = science and engineering; SDR = Survey of Doctorate Recipients; SED = Survey of Earned Doctorates; SEH = science, engineering, and health; SESTAT = Scientists and Engineers Statistical Data System.
SOURCE: Adapted from National Science Board (2016).

Appendix B

Questionnaire Content Crosswalk

National Survey of College Graduates, Survey of Doctorate Recipients, Survey of Earned Doctorates, and Early Career Doctorates Survey

Broad Concept	Specific Topic	NSCG/SDR Question Wording	SED Question Wording	ECDS Question Wording	Questionnaire Item Indicator			
					NSCG	SDR	SED	ECDS
Employment Situation	Working for pay or profit	**Were you working for pay or profit during the week of February 1, 2015?** *Working includes being self-employed, on a postdoctoral appointment, or on any type of paid or unpaid leave, including vacation.*	*Not collected*	**Have you worked for pay since receiving your doctorate degree?**	✓	✓		✓
	Looking for work	**Did you look for work at anytime during the four weeks preceding February 1, 2015? This would be between January 4th and February 1st.**	*Not collected*	*Not collected*	✓	✓		
	Reasons for not working	**What were your reasons for not working during the week of February 1, 2015?** (Mark Yes or No for each item) -Retired -On layoff from job -Student -Chronic illness or permanent disability -Suitable job not available -Did not need or want to work -Other	*Not collected*	**What were your reasons for not working for pay?** (Mark Yes or No for each item) -In school fulltime -In school parttime -Family responsibilities-child care -Family responsibilities-other -Chronic illness or permanent disability -Suitable position not available -Do not need or want to work -On layoff from a position -Retired -Other	✓	✓		✓
	Last worked date	**Prior to the week of February 1, 2015, when did you last work for pay or profit?** [Month and Year]	*Not collected*	*Not collected*	✓	✓		
	Last job-title	**What was the title of the last job you held prior to the week of February 1, 2015?**	*Not collected*	*Not collected*	✓	✓		
	Last job-job activities	**What kind of work were you doing on this last job -that is, what were your duties and responsibilities on your last job? Please be as specific as possible, including any areas of specialization.**	*Not collected*	*Not collected*	✓	✓		
	Last job-job category	**Using the JOB CATEGORY list on pages 18-19, choose the code that best describes the last job you held prior to the week of February 1, 2015.**	*Not collected*	*Not collected*	✓	✓		
	Previously retired indicator	**Although you were working during the week of February 1, 2015, had you previously retired from any position?**	*Not collected*	*Not collected*	✓	✓		

Broad Concept	Specific Topic	NSCG/SDR Question Wording	SED Question Wording	ECDS Question Wording	Questionnaire Item Indicator			
					NSCG	SDR	SED	ECDS
Employment Situation (continued)	Number of jobs since degree	*Not collected*	*Not collected*	**Since receiving your doctorate degree, how many different full-time paid positions and how many different part-time paid positions have you had, including your current position if applicable?**				✓
	Unpaid or volunteer position	*Not collected*	*Not collected*	**Have you held an unpaid or volunteer position since receiving your doctorate degree?**				✓
	Number of unpaid or volunteer positions	*Not collected*	*Not collected*	**Since receiving your doctorate degree, how many different unpaid positions have you had?** *Please include your current position if applicable.*				✓
Principal Employer	Employer information-name and location	**Who was your principal employer during the week of February 1, 2015?** (employer name, department/division, city, state, zip) *If you had more than one job, report the one for which you worked the most hours that week. If your employer had more than one location, report the location that employed you. If you worked for a contracting or consulting company, report the name of that company, not the client organization.*	*Not collected*	**Who is your employer for this position? What is the name and country of the place of your employment for this position?**	✓	✓		✓
	Employer information-main business or industry	**What was that employer's main business or industry; that is, what did that employer make or do?** *If your principal employer had more than one type of business, report the type of business primarily performed at the location where you worked.*	*Not collected*	*Not collected*	✓	✓		
	Employer information-number of employees	**Counting all locations where this employer operates, how many people worked for your principal employer? Your best estimate is fine.** (Mark one answer) -10 or fewer employees -11-24 employees -25-99 employees -100-499 employees -500-999 employees -1,000-4,999 employees -5,000-24,999 employees -25,000 or more employees	*Not collected*	*Not collected*	✓	✓		

Broad Concept	Specific Topic	NSCG/SDR Question Wording	SED Question Wording	ECDS Question Wording	Questionnaire Item Indicator			
					NSCG	SDR	SED	ECDS
Principal Employer (continued)	New business indicator	**Did your principal employer come into being as a new business within the past 5 years?**	*Not collected*	*Not collected*	✓	✓		
	Employer type	**Which of the following best describes your principal employer during the week of February 1, 2015? Were you...** (Mark one answer) -SELF-EMPLOYED in a nonincorporated business, professional practice, or farm -SELF-EMPLOYED in an incorporated business, professional practice, or farm -PRIVATE SECTOR employee in a for-profit company or organization -PRIVATE SECTOR employee in a nonprofit organization -GOVERNMENT employee in a local government -GOVERNMENT employee in a state government -GOVERNMENT employee in the U.S. military service active duty or Commissioned Corps -GOVERNMENT employee in the U.S. government (e.g., civilian employee) -Other type of employee	*Not collected*	**Which one of the following best describes the type of organization where this position is located?** - Public or private educational institution including university-affiliated research institutes, hospitals, and medical centers -Government including military and intergovernmental organizations -Federally Funded Research and Development Center (FFRDC) -Hospital, medical care facility, or community health system not affiliated with a university -Other for-profit business -Other non-profit organization or foundation -Other	✓	✓		✓
		Was your principal employer an educational institution?	*Not collected*		✓	✓		✓
		Was the educational institution where you worked a... (Mark one answer) -Preschool, elementary, middle, or secondary school -2-year college, community college, or technical institute -4-year college or university, other than a medical school -Medical school (including university-affiliated hospital) -University-affiliated research institute	*Not collected*	(*If an educational institution*) **Which one of the following best describes the type of educational institution where this position is located?** -Medical school including university-affiliated hospital or medical center -University or a 4-year college other than a medical school -University-affiliated research institute -Community or 2-year college -Preschool, elementary, middle, secondary school or school system	✓	✓		✓

Broad Concept	Specific Topic	NSCG/SDR Question Wording	SED Question Wording	ECDS Question Wording	Questionnaire Item Indicator			
					NSCG	SDR	SED	ECDS
Principal Employer (continued)	Employer type (continued)	*Not collected*	*Not collected*	(*If a government institution*) **Which one of the following best describes the type of government institution where this position [is/was] located?** -U.S. military -U.S. federal government (non-military) -U.S. state government -U.S. local government -Foreign government or military -Intergovernmental organization				✓
		Not collected	*Not collected*	(*If a non-profit*) **Is this non-profit organization a...** -Research institute, including medical research organization -Hospital, medical care facility, or community health system not affiliated with a university -University-affiliated hospital or medical center -Private foundation -Professional or technical society -Advocacy group -Other				✓
Principal Job-Academic Position	Academic position	**During the week of February 1, 2015, what type of academic position(s) did you hold at this institution?** (Mark Yes or No for each item) -President, Provost, or Chancellor (any level) -Dean (any level), department head, or department chair -Research faculty, scientist, associate, or fellow -Teaching faculty -Adjunct faculty -Postdoc (e.g., postdoctoral fellow or associate) -Research assistant -Teaching assistant -Other position	*Not collected*	**Do you hold any of the following academic positions?** -Research faculty, scientist, associate, or fellow -Teaching faculty -Adjunct faculty -Postdoc (e.g., postdoctoral fellow or associate) -Research assistant -Teaching assistant -President, Provost, or Chancellor (any level) -Dean (any level), department head -Other		✓		✓

Broad Concept	Specific Topic	NSCG/SDR Question Wording	SED Question Wording	ECDS Question Wording	Questionnaire Item Indicator			
					NSCG	SDR	SED	ECDS
Principal Job-Academic Position (continued)	Faculty rank	**What was your faculty rank?** (Mark one answer) -Not applicable: no ranks designated at this institution -Not applicable: no ranks designated for my position -Professor -Associate Professor -Assistant Professor -Instructor -Lecturer -Other	*Not collected*	**What is your faculty rank?** -Professor -Associate Professor -Assistant Professor -Instructor -Lecturer -Not applicable: no ranks designated at your institution -Not applicable: no ranks designated for your position -Other		✓		✓
	Tenure status	**What was your tenure status?** (Mark one answer) -Not applicable: no tenure system at this institution -Not applicable: no tenure system for my position -Tenured (and year) -On tenure track, but not tenured -Not on tenure track	*Not collected*	**What is your tenure status?** -Not applicable: No tenure system at your institution -Not applicable: No tenure system for your position -Tenured -On tenure track but not tenured -Not on tenure track		✓		✓
		What year were you tenured? (Year tenured, for those who Responded "Tenured")	*Not collected*	*Not collected*		✓		
	Postdoc status	**Was this job a "postdoc"?** *A "postdoc" is a temporary position awarded in academe, industry, a non-profit organization, or government primarily for gaining additional education and training in research.*	*Not collected*	**Would you consider your first full-time paid position following the receipt of your doctorate degree a postdoctoral appointment?** **Would your organization consider this position a postdoctoral appointment?**		✓		✓
		What were your reasons for taking this postdoc? (Mark Yes or No for each item) -Additional training in PhD field -Training in an area outside of PhD filed -Work with a specific or in a specific place -Other employment not available -Postdoc generally expected for a career in this field -Some other reason	*Not collected*	*Not collected*		✓		
		Which two reasons in the previous question were your most important reasons for taking this postdoc?	*Not collected*	*Not collected*		✓		

Broad Concept	Specific Topic	NSCG/SDR Question Wording	SED Question Wording	ECDS Question Wording	Questionnaire Item Indicator			
					NSCG	SDR	SED	ECDS
Principal Job-Academic Position (continued)	Postdoc status (continued)	*Not collected*	*Not collected*	**How important is a postdoctoral appointment to advancement in your desired career track within your specific field?**				✓
		Not collected	*Not collected*	**How many postdoctoral appointments have you had, including your current position if applicable?**				✓
		Not collected	*Not collected*	**Did your organization have...** -a postdoctoral association? -a postdoctoral office? -a postdoctoral related agreement (i.e., a "Compact" or "Memorandum of Understanding")?				✓
	Clinical position	*Not collected*	*Not collected*	**Were you a...** -clinical researcher? -clinical fellow? -clinical practitioner?				✓
Principal Job	Principal job-title	**What was the title of the principal job you held during the week of February 1, 2015?**	*Not collected*	**What is the official title of this position?**	✓	✓		✓
	Principal job-job activities	**What kind of work were you doing on this job-that is, what were your duties and responsibilities on your principal job? Please be as specific as possible, including any areas of specialization.**	*Not collected*	*Not collected*	✓	✓		
	Principal job-job category	**Using the JOB CATEGORY list on pages 18-19, choose the code that best describes the principal job during the week of February 1, 2015.**	*Not collected*	**In what field are you working?**	✓	✓		✓
	Principal job requirements	**Did your duties on this job require technical expertise of a bachelor's degree in...** (Mark Yes or No for each item) -engineering, computer science, math, or the natural sciences? -the social sciences? -some other field (e.g., health, business, or education)?	*Not collected*	*Not collected*	✓	✓		
	Principal job start date	**During what month and year did you start this job, (i.e., your principal job held during the week of February 1, 2015)?** [month and year response]	*Not collected*	**In what month and year did your first full-time paid position following the receipt of your doctorate degree begin?** **If this position has ended, in what month and year did it end?**	✓	✓		✓

Broad Concept	Specific Topic	NSCG/SDR Question Wording	SED Question Wording	ECDS Question Wording	Questionnaire Item Indicator			
					NSCG	SDR	SED	ECDS
Principal Job (continued)	Relationship of education and job	NSCG: **To what extent was your work on your principal job related to your highest degree? Was it...** SDR: **To what extent was your work on your principal job related to your first U.S. doctoral degree? Was it...** -closely related -somewhat related -not related	*Not collected*	**To what extent is this position related to your doctoral degree? Would you say...** -closely related -somewhat related -not related	✓	✓		✓
		NSCG: *(If job and degree were not related)* **Did these factors influence your decision to work in an area outside the field of your highest degree?** SDR: (If job and degree were not related) **Did these factors influence your decision to work in an area outside the field of your first U.S. doctoral degree?** -Pay, promotion opportunities -Working conditions -Job location -Change in career or professional interests -Family-related reasons -Job in highest degree field not available -Some other reason	*Not collected*	(If job and degree were not related) **Did any of the following factors influence your decision to work in an area outside the field of your doctoral degree?** -Pay, promotion opportunities -Working conditions -Personal career or professional interests -Family-related reasons -Geographic location of position -Position in your doctorate field of study was not available -Opportunity for training outside doctorate field of study -Some other reason	✓	✓		✓
		NSCG: **Which two factors are your most important reasons that influence your decision to work in an area outside the field of your highest degree?** SDR: **Which two factors are your most important reasons that influence your decision to work in an area outside the field of your first U.S. doctoral degree?**	*Not collected*	*Not collected*	✓	✓		

Broad Concept	Specific Topic	NSCG/SDR Question Wording	SED Question Wording	ECDS Question Wording	Questionnaire Item Indicator			
					NSCG	SDR	SED	ECDS
Principal Job (continued)	Job characteristics	*Not collected*	*Not collected*	**For each of the following job characteristics, please indicate if it describes your position:** -Requires a PhD or doctorate-equivalent degree (e.g., MD, DO, DDS, DVM) -Requires that the doctorate was awarded within the last 10 years -Is a temporary position -Is intended to provide training in research -Is intended to advance professional skills -Requires a full-time commitment to research -Requires that you work under the direction of a senior scholar or principal investigator -Is intended to be for a defined period of time (i.e., [has/had] a term limit) -Requires publication of research in scholarly journals -Requires mentorship for professional development -Is intended to prepare you for an independent career -Has an expectation that you provide service to your institution, profession, or community (e.g., committee work, serving as a reviewer)				✓

Broad Concept	Specific Topic	NSCG/SDR Question Wording	SED Question Wording	ECDS Question Wording	Questionnaire Item Indicator			
					NSCG	SDR	SED	ECDS
Principal Job (continued)	Work activities	**Which of the following work activities occupied at least 10 percent of your time during a typical work week on this job?** (Mark Yes or No for each item) -Accounting, finance, contracts -Basic research -Applied research -Development -Design of equipment, processes, structures, models -Computer programming, systems or applications development -Human resources -Managing or supervising people or projects -Production, operations, maintenance -Professional services -Sales, purchasing, marketing, customer service, public relations -Quality or productivity management -Teaching -Other	*Not collected*	**The next questions ask about your work activities in this position. While in this position have you...** (Mark Yes or No for each item) -written technical reports or working papers other than those presented at conferences or submitted for publication? -attended professional conferences when you were not presenting papers? -served as a reviewer for a conference, journal, or an agency, institution, or organization? -supervised graduate students, postdocs, or research staff? -taught courses? -developed marketable products? -worked with researchers not a part of your department or lab? -served as the principal or co-investigator on funded grants? -prepared grant or fellowship proposals in which you are listed as the principal or co- investigator? -prepared grant proposals in which you are <u>not</u> listed as the principal or co-investigator?	✓	✓		✓
		On which <u>two</u> work activities in the previous question did you work the <u>most</u> hours during a typical week on this job?	*Not collected*	**On which <u>two</u> work activities in the previous question did you work the <u>most</u> hours during a typical week on this job?**	✓	✓		✓
	Supervision responsibilities	**Did you supervise the work of others as part of the principal job you held during the week of February 1, 2015?**	*Not collected*	*Not collected*	✓	✓		
		How many people did you typically supervise directly?	*Not collected*	*Not collected*	✓	✓		
		How many people did you typically supervise indirectly through subordinate supervisors?	*Not collected*	*Not collected*	✓	✓		

Broad Concept	Specific Topic	NSCG/SDR Question Wording	SED Question Wording	ECDS Question Wording	Questionnaire Item Indicator			
					NSCG	SDR	SED	ECDS
Principal Job (continued)	Performance evaluation	*Not collected*	*Not collected*	**How often do you and your supervisor meet to discuss your work?** -Once a day or more -At least once a week but not every day -At least once a month but not every week -At least once a year but not every month -Never -Do not have a supervisor				✓
		Not collected	*Not collected*	**Did you and your supervisor write a plan that outlines the expectations and responsibilities of your position?**				✓
		Not collected	*Not collected*	(*If no performance plan exists*) **Have you and your supervisor discussed the option of developing a plan?**				✓
		Not collected	*Not collected*	(If a performance plan does exist) **How much does the plan reflect your personal career development goals? Would you say…** -not at all -somewhat -a great deal				✓
		Not collected	*Not collected*	**Has your supervisor provided you with a formal performance evaluation?** **If no, do you expect a performance evaluation in the future?**				✓

Broad Concept	Specific Topic	NSCG/SDR Question Wording	SED Question Wording	ECDS Question Wording	Questionnaire Item Indicator			
					NSCG	SDR	SED	ECDS
Principal Job (continued)	Performance evaluation (continued)	*Not collected*	*Not collected*	**With regard to your supervisor, please indicate the extent to which you agree or disagree with the following statements. Your supervisor...** -provides appropriate recognition for your work -encourages collaboration with other researchers -encourages increasing independence in your research -encourages presenting work outside the department or lab -provides opportunities to work on grant proposals -supports publishing papers -provides career guidance -encourages work-life balance				✓
	Job satisfaction	**Thinking about your principal job held during the week of February 1, 2015, please rate your satisfaction with that job's ...** (Mark one answer for each item. Response options are very satisfied, somewhat satisfied, somewhat dissatisfied, very dissatisfied) -salary -benefits -job security -job location -opportunities for advancement -intellectual challenge -level of responsibility -degree of independence -contribution to society	*Not collected*	*Not collected*	✓	✓		
		How would you rate your overall satisfaction with the principal job you held during the week of February 1, 2015? -Very satisfied -Somewhat satisfied -Somewhat dissatisfied -Very dissatisfied	*Not collected*	**How satisfied are you with this position? Would you say...** -Very dissatisfied -Dissatisfied -Neither dissatisfied nor satisfied -Satisfied -Very satisfied	✓	✓		✓

Broad Concept	Specific Topic	NSCG/SDR Question Wording	SED Question Wording	ECDS Question Wording	Questionnaire Item Indicator			
					NSCG	SDR	SED	ECDS
Principal Job (continued)	Job satisfaction (continued)	*Not collected*	*Not collected*	**How much does this position meet your expectations in the following areas?** -Increase in subject matter knowledge -Conduct your own independent research -Publish papers -Present papers at professional conferences -Teach -Gain supervisory experience -Write grant proposals -Grant submissions under your name -Gain experience in an interdisciplinary research environment -Network with colleagues -Develop marketable products or technologies				✓
	Starting salary	*Not collected*	*Not collected*	**What was your basic annual salary when you started the position?**				✓
	Current salary	**As of the week of February 1, 2015, what was your basic annual salary on your principal job, before deductions?**	*Not collected*	**What is your basic annual salary before deductions for this position?**	✓	✓		✓
	Salary disbursing	*Not collected*	*Not collected*	**Over how many months was your annual salary disbursed?**				✓
	Number of weeks worked per year	**Was this salary based on a 52-week year, or less than that?** If less than 52 weeks, enter the number of weeks per year.	*Not collected*	*Not collected*	✓	✓		
	Hours worked	**During a typical week on your principal job, how many hours did you work?**	*Not collected*	**During a typical week in this position, how many hours do you work?**	✓	✓		✓
		(If fewer than 35 hours) **Did you want to work 35 or more hours per week on your principal job?**	*Not collected*	*Not collected*	✓	✓		
		For which of the following reasons did you usually work fewer than 35 hours per week on the principal job you held during the week of February 1, 2015? (Mark Yes or No for each item) -Previously retired or semi-retired -Student -Family responsibilities -Full-time job not available -Held more than one job -Did not need or want to work more hours -Other	*Not collected*	*Not collected*	✓	✓		

Broad Concept	Specific Topic	NSCG/SDR Question Wording	SED Question Wording	ECDS Question Wording	Questionnaire Item Indicator			
					NSCG	SDR	SED	ECDS
Principal Job (continued)	Job benefits	**Concerning your principal job during the week of February 1, 2015, were any of the following benefits available to you, even if you chose not to take them?** (Mark Yes or No for each item) - Health insurance that was at least partially paid by your employer - A pension plan or a retirement plan to which your employer contributed - A profit-sharing plan - Paid vacation, sick, or personal days	*Not collected*	**Thinking about this position, do you receive any of the following benefits, even if you chose not to take them?** (Mark Yes or No for each item) - Dental insurance - Vision insurance - Health insurance - Paid sick days - Paid vacation days - Paternity or maternity leave - A retirement plan or pension - Childcare subsidy - Tuition reimbursement - Other	✓	✓		✓
	Funding source for position	*Not collected*	*Not collected*	**Which of the following sources fund this position?** - U.S. federal research grant or contract - U.S. federal fellowship or training grant - U.S. state or public university funds - Private university funds - U.S. city or local government funds - Foreign (non-U.S.) government funds - Non-profit organization or foundation funds - For-profit business funds - Personal or other individually held funds - Other				✓
	Federal support for work indicator	**Thinking back now to 2014, was any of your work during 2014 supported by contracts or grants from the U.S. government?**	*Not collected*	**How many U.S. federal grants and contracts support this position?** - One - More than one (Please specify number) - Don't know	✓	✓		✓
	Federal support for work-funding recipient	*Not collected*	*Not collected*	**Thinking about the U.S. federal grants or contracts that fund this position, to whom were these awarded?** - Directly to you - Your supervisor or the Principal Investigator of the project you work on - A research center, consortium, or lab - Other				✓

Broad Concept	Specific Topic	NSCG/SDR Question Wording	SED Question Wording	ECDS Question Wording	Questionnaire Item Indicator			
					NSCG	SDR	SED	ECDS
Principal Job (continued)	Federal support for work-funding agency	**Which federal agencies or departments were supporting your work?** (Mark all that apply) -Department of Defense (DOD) -Department of Education -Department of Energy (DOE) -Department of Health and Human Services (except NIH) -National Aeronautics and Space Administration (NASA) -National Institutes of Health (NIH) -National Science Foundation (NSF) -Other -Don't know source agency	*Not collected*	**Which government agencies funded this position?** -Department of Agriculture (USDA) -Department of Commerce (Commerce) -Department of Defense (DOD) -Department of Education (ED) -Department of Energy (DOE) -Department of Homeland Security (DHS) -Department of the Interior (DOI) -Department of Labor (DOL) -Department of Transportation (DOT) -Department of Veterans Affairs (VA) -National Aeronautics and Space Administration (NASA) -National Institutes of Health (NIH) -National Science Foundation (NSF) -Other -Don't know	✓	✓		✓
	Earned income	**<u>Counting all jobs held</u> in 2014, what was your total earned income for 2014, <u>before</u> deductions?** *Include all wages, salaries, bonuses, overtime, commissions, consulting fees, net income from businesses, summertime teaching or research, or other work associated with scholarships.*	*Not collected*	*Not collected*	✓	✓		
	Position search tools	*Not collected*	*Not collected*	**What was the primary way you found this position?** -An advertisement in a journal, publication, or website -Through your graduate advisor -Through a mentor other than your graduate advisor -Through another colleague -Directly contacted by this employer other than by mentor or colleague -You directly contacted this employer -You were promoted from an internship within the organization -You were promoted from another position within the organization -Career or job fair -Professional recruiter -Other				✓

Broad Concept	Specific Topic	NSCG/SDR Question Wording	SED Question Wording	ECDS Question Wording	Questionnaire Item Indicator			
					NSCG	SDR	SED	ECDS
Job Search for Principal Job	Other employment options indicator	*Not collected*	*Not collected*	**Were you offered other positions at the time you accepted this position?**				✓
		Not collected	*Not collected*	**What type of positions were you offered at the time you accepted this position? Were you offered...** -permanent position(s) -postdoc(s) -temporary position other than a postdoc				✓
	Factors that influenced job selection	*Not collected*	*Not collected*	**What factors did you consider in your decision to choose this position?** -It was the only position you were offered at the time -Opportunity to publish -Opportunity to conduct your own independent research -Opportunity to increase knowledge in your doctorate degree field of study -Opportunity for training outside of your doctorate degree field of study -Opportunity to work with a particular researcher -Likelihood of receiving outside funding -Success of this supervisor's former staff (e.g., postdocs, research staff) -Prestige of the institution -Opportunity to work in a particular lab -Promotion opportunities -Salary -Benefits provided by the institution (e.g., health insurance) -Family-related reasons (e.g., children, spouse's or partner's job location) -Geographic location of position -Working conditions (e.g., hours, equipment, and working environment) -Other				✓
		Not collected	*Not collected*	**Which factor was most important in your decision to choose this position?**				✓

Broad Concept	Specific Topic	NSCG/SDR Question Wording	SED Question Wording	ECDS Question Wording	Questionnaire Item Indicator			
					NSCG	SDR	SED	ECDS
Certifications and Licenses (continued)	Certification field of study	**What is the primary subject or field of study for this certification or license?**	*Not collected*	*Not collected*	✓			
	Certification issuer	**Who issued this certification or license?** (Mark one answer) -Federal, state, or local government -Professional or trade association -Business or company -Other group or organization	*Not collected*	*Not collected*	✓			
	Reasons for holding a certification	**Why do you currently hold this certification or license?** (Mark Yes or No for each item) -To improve skills or knowledge in your current occupational field -To increase opportunities for promotion or advancement in your current occupational field -To facilitate a change to a different occupational field -Required or expected by employer -To start your own business -Other	*Not collected*	*Not collected*	✓			
	Certification-CEU indicator	**Do you have to earn continuing education units (CEUs) or other professional development credits to maintain this certification or license?**	*Not collected*	*Not collected*	✓			
	Certification funding	**Were any of your costs for this certification or license paid for by an employer?**	*Not collected*	*Not collected*	✓			
	Certification-work-related indicator	**Is this certification or license for the principal job you help during the week of February 1, 2015?** *If you were not working, please answer No.*	*Not collected*	*Not collected*	✓			
Past Employment	Continued employment indicator	**Were you working for pay or profit during both of these time periods—the week of February 1, 2013, and the week of February 1, 2015?**	*Not collected*	*Not collected*	✓	✓		

Broad Concept	Specific Topic	NSCG/SDR Question Wording	SED Question Wording	ECDS Question Wording	Questionnaire Item Indicator			
					NSCG	SDR	SED	ECDS
Past Employment (continued)	Employer/job change indicator	**During these two time periods—the week of February 1, 2013, and the week of February 1, 2015— were you working for…** -same employer and in same type of job -same employer but in same type of job -different employer but in same type of job -different employer and in different type of job	*Not collected*	*Not collected*	✓	✓		
	Reasons for changing employer or job	*(if different)* **Why did you change your employer or your job?** (Mark Yes or No for each item) -Pay, promotion opportunities -Working conditions (e.g., hours, equipment, working environment) -Job location -Change in career or professional interests -Family-related reasons (e.g., children, spouse's job moved) -School-related reasons (e.g., returned to school, completed a degree) -Laid off or job terminated (includes company closings, mergers, buyouts, grant or contract ended) -Retired	*Not collected*	*Not collected*	✓	✓		

Broad Concept	Specific Topic	NSCG/SDR Question Wording	SED Question Wording	ECDS Question Wording	Questionnaire Item Indicator			
					NSCG	SDR	SED	ECDS
Professional Activities	Type of activities	*Not collected*	*Not collected*	**Thinking about all of your paid, unpaid, and other research activities since receiving your degree, have you...** -published papers in conference proceedings? -(co)authored papers for presentation at regional, national, or international conferences? -had articles accepted for publication or already published in a peer-reviewed journal? -submitted articles for publication in a peer-reviewed journal that were not accepted for publication or published? -published books? -published book chapters? -been named as an inventor on U.S. patent applications? -been named as an inventor on foreign patent applications?				✓
	Number of activities	*Not collected*	*Not collected*	**How many...** -papers have you published in conference proceedings? -papers have you (co)authored for presentation at regional, national, or international conferences? (do not include papers published in conference proceedings.) -articles have you had accepted for publication or already published in a peer-reviewed journal? -articles have you submitted for publication in a peer-reviewed journal that were not accepted for publication or published? (If you have submitted one paper multiple times, count it only once.) -books have you published? -book chapters have you published? -U.S. patent applications have you been named as an inventor on? -foreign patent applications have you been named as an inventor on?				✓

Broad Concept	Specific Topic	NSCG/SDR Question Wording	SED Question Wording	ECDS Question Wording	Questionnaire Item Indicator			
					NSCG	SDR	SED	ECDS
Professional Activities (continued)	Patents granted	*Not collected*	*Not collected*	**How many patents have been granted to you as an inventor?**				✓
	Patents purpose	*Not collected*	*Not collected*	**Thinking of the U.S. patents that you have been granted, for which of the following types of activity have they been granted?** -utility patents -plant patents -design patents				✓
	Outcome of patents	*Not collected*	*Not collected*	**Have any of your patents resulted in commercialized products or processes or been licensed?**				✓
Other Work-related Experience	Work-related training indicator	**During the past 12 months, did you attend any work-related training, such as workshops or seminars?** *Include conferences or professional meetings only if you attended a training session at the conference or meeting. Do not include college coursework for which you were enrolled in a degree program.*	*Not collected*	**During the past 12 months, did you attend any work-related training, such as workshops or seminars?** *Include conferences or professional meetings only if you attended a training session at the conference or meeting. Do not include college coursework for which you were enrolled in a degree program.*	✓	✓		✓
	Reasons for work-related training	*(If Yes)* **For which of the following reasons did you take training during the past 12 months?** (Mark Yes or No for each item) -To improve skills or knowledge in your current occupational field -To increase opportunities for promotion or advancement in your current occupational field -For licensure or certification in your current occupational field -To facilitate a change to a different occupational field -Required or expected by employer -For leisure or personal interest -Other	*Not collected*	*(If Yes)* **For which of the following reasons did you take training during the past 12 months?** (Mark Yes or No for each item) -To improve skills or knowledge in your current occupational field -To increase opportunities for promotion or advancement in your current occupational field -For licensure or certification in your current occupational field -To facilitate a change to a different occupational field -Required or expected by employer -For leisure or personal interest -Other	✓	✓		✓
		What was your most important reason from the previous question for taking training?	*Not collected*	*Not collected*	✓	✓		

Broad Concept	Specific Topic	NSCG/SDR Question Wording	SED Question Wording	ECDS Question Wording	Questionnaire Item Indicator			
					NSCG	SDR	SED	ECDS
Other Work-related Experience (continued)	Work-related training topics	*Not collected*	*Not collected*	**While in this position, did you attend any training workshops or training courses about the following topics?** -Grant writing -Publishing -Orientation to various career paths -Supervising people -Project management -Teaching -Intellectual property -Research ethics -Other training workshops or training courses				✓
	Professional associations	**During the past 12 months, did you attend any professional society or association meetings or professional conferences?** *Include regional, national, or international meetings*	*Not collected*	*Not collected*	✓	✓		
		To how many regional, national, or international professional societies or associations do you currently belong?	*Not collected*	*Not collected*	✓	✓		
	Importance of job characteristics	**When thinking about a job, how important is each of the following factors to you?** (Mark one answer for each item. Response options are very satisfied, somewhat satisfied, somewhat dissatisfied, very dissatisfied) -Salary -Benefits -Job security -Job location -Opportunities for advancement -Intellectual challenge -Level of responsibility -Degree of independence -Contribution to society	*Not collected*	*Not collected*	✓	✓		

Broad Concept	Specific Topic	NSCG/SDR Question Wording	SED Question Wording	ECDS Question Wording	Questionnaire Item Indicator			
					NSCG	SDR	SED	ECDS
Work-Life Balance	Postgraduate period	*Not collected*	*Not collected*	**What were you doing between the date your degree was awarded and the date you began full-time employment?** -Working in a part-time paid position -Working in an unpaid position either fulltime or parttime -Not working -Other				✓
	Time away from work indicator	*Not collected*	*Not collected*	**Since you started your first full-time paid position, have you ever taken time off for 6 months or more?**				✓
	Reasons for time away from work	*Not collected*	*Not collected*	*(If Yes)* **Why did you take time off?** -For health reasons -To start a family or have more children -To care for a sick family member -To write papers from your dissertation for publication -To travel -For additional education -You could not find employment -For some other reason				✓
	Time away from work	*Not collected*	*Not collected*	*(If Yes)* **How many months did you take off?**				✓
	Work stress level	*Not collected*	*Not collected*	**Would you say that your position is…** -not stressful at all -somewhat stressful -very stressful -extremely stressful				✓

Broad Concept	Specific Topic	NSCG/SDR Question Wording	SED Question Wording	ECDS Question Wording	Questionnaire Item Indicator			
					NSCG	SDR	SED	ECDS
Work-Life Balance (continued)	Balancing priorities	*Not collected*	*Not collected*	**Please indicate how much you agree or disagree with the following statements about balancing your professional and personal responsibilities.** -You can manage the demands of your position and at home -Demands at home limit your professional activities (e.g., going to conferences) -Your work schedule allows you to maintain the overall quality of life you want -Your work schedule provides the flexibility to take care of demands at home -Your supervisor understands when demands at home interfere with your professional responsibilities -Demands at home have slowed down progress on your professional activities				✓
Education Background	High school year	**In what year did you receive your high school diploma or high school equivalency certificate?**	*Not collected*	*Not collected*	✓			
	High school location	**In what U.S. state, U.S. territory, or foreign country did you last attend high school?**	**In what state or country was the high school/secondary school that you last attended?**	*Not collected*	✓		✓	
	Community college	**Have you <u>ever</u> taken courses at a community college?**	**Did you earn college credit from a community or two-year college?**	**Have you ever taken a course at a community college?**	✓		✓	✓
	Community college time period	*(If Yes)* **Did you take courses at a community college during any of the following time periods?** (Mark Yes or No for each item) -Before graduating from high school or earning a high school equivalency certificate -After high school and before ever enrolling in a 4-year college or university -While enrolled in a 4-year college or universityand before receiving your first bachelor's degree -After leaving a 4-year college or university without receiving your first bachelor's degree -Any time after receiving your first bachelor's degree	*Not collected*	*Not collected*	✓			

Broad Concept	Specific Topic	NSCG/SDR Question Wording	SED Question Wording	ECDS Question Wording	Questionnaire Item Indicator			
					NSCG	SDR	SED	ECDS
Education Background (continued)	Reasons for community college attendance	**Thinking back to the time(s) you attended community college, for which of the following <u>reasons</u> did you take community college courses?** (Mark Yes or No for each item) -To earn college credits while still attending high school -To complete an associate's degree -To prepare for college/increase chance of acceptance to a 4-year college or university -To earn credits for a bachelor's degree -For financial reasons (e.g., cost of a 4-year school) -To gain further skills or knowledge in your academic or occupational field -To facilitate a change in your academic or occupational field -To increase opportunities for promotion, advancement, or higher salary -Other	*Not collected*	*Not collected*	✓			
		Which <u>two</u> reasons in the previous question were your <u>most important</u> reasons for taking community college courses?	*Not collected*	*Not collected*	✓			
	Associate's degree	**Do you have an associate's degree?**	*Not collected*	*Not collected*	✓			
	Associate's degree-school name	*(If Yes)* **From which academic institution did you receive your associate's degree?** (school name, department, city, state)	*Not collected*	*Not collected*	✓			
	Bachelor's degree	**Do you have a degree at the bachelor's level or higher?**	*Not collected*	*Not collected*	✓			
	Number of degrees	**How many college or university degrees did you receive at the bachelor's level or higher prior to February 1, 2015?**	*Not collected*	**Not including your doctorate degree, how many other graduate, professional, or undergraduate degrees have you received?**	✓			✓
	MD or DDS indicator	*Not collected*	**Are you earning, or have you earned, an MD or a DDS?**	*Not collected*			✓	
Doctoral Program Information	Dissertation title	*Not collected*	**What is the title of your dissertation?**	*Not collected*			✓	
	Dissertation primary field of research	*Not collected*	**Please write the name of the primary field of your dissertation research.**	*Not collected*			✓	

Broad Concept	Specific Topic	NSCG/SDR Question Wording	SED Question Wording	ECDS Question Wording	Questionnaire Item Indicator			
					NSCG	SDR	SED	ECDS
Doctoral Program Information (continued)	Dissertation primary field of research code	*Not collected*	**Using the list on pages 6-7, choose the code that best describes the primary field of your dissertation research.**	*Not collected*			✓	
	Dissertation secondary field of research	*Not collected*	**If your dissertation research was interdisciplinary, list the number of your secondary field.** (Respondent can provide up to three secondary fields)	*Not collected*			✓	
	Dissertation secondary field of research code	*Not collected*	**If your dissertation research was interdisciplinary, list the number of your secondary field.** (Respondent can provide up to three secondary fields)	*Not collected*			✓	
	Department name	*Not collected*	**Please name the department (or interdisciplinary committee, center, institute, etc.) of the university that supervised your doctoral studies.**	*Not collected*			✓	
	Prerequisites	*Not collected*	**Was a master's degree a prerequisite for admission to your doctoral program?**	*Not collected*			✓	
	Initial graduate school enrollment	*Not collected*	**In what month and year did you first enter any graduate school in any program or capacity?**	*Not collected*			✓	
	Coursework timing	*Not collected*	**How many years were you taking courses or preparing for exams for this doctoral degree (including a master's degree, if that was part of your doctoral program)?**	*Not collected*			✓	
	Dissertation timing	*Not collected*	**How many years were you working on your dissertation after coursework and exams (non-course related preparation and research, writing, and defense)?**	*Not collected*			✓	
	Dissertation break indicator	*Not collected*	**Was there any time between the year you entered your doctoral program and the award of your doctorate that you were not working on your degree (i.e., not taking courses or working on your dissertation)?** **If yes, how many years were you not working on your degree?**	*Not collected*			✓	
Most Recent Degree	School name	**From which school did you receive your most recent degree?** (name, department, city, state)	**Institution name** (name, branch or city, state or province, country)	**What is the full name and location of the institution from which you received this degree?** (name, branch or city, state, country)	✓		✓	✓
	Degree completion date	**In what month and year was this degree awarded?**	**Month/year degree granted**	**In what month and year was this degree awarded?**	✓		✓	✓
	Degree start date	*Not collected*	**Month/year degree that you started your degree**	*Not collected*			✓	

Broad Concept	Specific Topic	NSCG/SDR Question Wording	SED Question Wording	ECDS Question Wording	Questionnaire Item Indicator			
					NSCG	SDR	SED	ECDS
Most Recent Degree (continued)	Degree level	**What type of degree did you receive?** -Bachelor's degree (e.g., BS, BA, AB) -Master's degree (e.g., MS, MA, MBA) -Doctorate (e.g., PhD, DSc, EdD) -Other professional degree (e.g., JD, LLB, MD, DDS, DVM) -Other	**Have you received a degree of this type?** -Bachelor's degree (e.g., BA, BS, AB) or equivalent -Master's degree (e.g., MS, MA, MBA) or equivalent -Research doctorate degree -Other	**Thinking about your highest degree, what type of degree is it?** *(If you have more than one doctorate or doctorate-equivalent degree, think about the first such degree you received.*) -Doctorate degree (e.g., PhD, EdD, DSc, Candidate of science, Dr. rer. nat., etc.) -Doctorate-equivalent degree (e.g., MD, DO, DDS, DVM, etc.) -Master's degree (e.g., MS, MA, MPH, Diploma, etc.) -Undergraduate degree (e.g., BS, BA, AB)	✓		✓	✓
	Doctorate degree type	*Not collected*	*Not collected*	(*If degree level is doctorate degree*) **What type of doctorate degree is it?** -PhD (Doctor of Philosophy) -EdD (Doctor of Education) -DSc (Doctor of Science) -Candidate of science -Dr. rer. nat. -Other				✓
	Doctorate equivalent degree type	*Not collected*	*Not collected*	(*If degree level is doctorate-equivalent degree*) **What type of doctorate-equivalent degree is it?** -DDS (Doctor of Dental Surgery) -DO (Doctor of Osteopathic Medicine) -DVM (Doctor of Veterinary Medicine) -MD (Doctor of Medicine) -Other				✓
	International equivalent indicator	*Not collected*	*Not collected*	**Is your highest degree an international equivalent to a U.S. doctorate or doctorate-equivalent degree?**				✓
	Major field of study and second major	**What is the major field of study and second major (if any) for this degree?**	**Primary field of study**	**What was your primary field of study for this degree?**	✓		✓	✓
	Major category code	**Using the FIELD OF STUDY list on pages 20-21, choose the code that <u>best</u> describes the major field of study and second major (if any) for this degree**	**Field number form list on pp. 6-7**	*Not collected*	✓		✓	
Second Most Recent Degree	School name	**From which school did you receive your second most recent degree?** (name, department, city, state)	**Institution name** (name, branch or city, state or province, country)	**What is the full name and location of the institution from which you received this degree?** (name,bBranch orcCity, state, country)	✓		✓	✓

Broad Concept	Specific Topic	NSCG/SDR Question Wording	SED Question Wording	ECDS Question Wording	Questionnaire Item Indicator			
					NSCG	SDR	SED	ECDS
Second Most Recent Degree (continued)	Degree completion	**In what month and year was this degree awarded?**	**Month/year degree granted**	**In what month and year was this degree awarded?**	✓		✓	✓
	Degree start date	*Not collected*	**Month/year degree that you started your degree**	*Not collected*			✓	
	Degree level	**What type of degree did you receive?** -Bachelor's degree (e.g., BS, BA, AB) -Master's degree (e.g., MS, MA, MBA) -Doctorate (e.g., PhD, DSc, EdD) -Other professional degree (e.g., JD, LLB, MD, DDS, DVM) -Other	**Have you received a degree of this type?** -Bachelor's degree (e.g., BA, BS, AB) or equivalent -Master's degree (e.g., MS, MA, MBA) or equivalent -Research doctorate degree -Other	**Thinking about your highest degree, what type of degree is it?** *(If you have more than one doctorate or doctorate-equivalent degree, think about the first such degree you received.)* -Doctorate degree (e.g., PhD, EdD, DSc, Candidate of science, Dr. rer. nat., etc.) -Doctorate-equivalent degree (e.g., MD, DO, DDS, DVM, etc.) -Master's degree (e.g., MS, MA, MPH, Diploma, etc.) -Undergraduate degree (e.g., BS, BA, AB)	✓		✓	✓
	Major field of study and second major	**What is the major field of study and second major (if any) for this degree?**	**Primary field of study**	**What was your primary field of study for this degree?**	✓		✓	✓
	Major category code	**Using the FIELD OF STUDY list on pages 20-21, choose the code that best describes the major field of study and second major (if any) for this degree)**	**Field number form list on pages 6-7**	*Not collected*	✓		✓	
First Bachelor's Degree	School name	**From which school did you receive your first bachelor's degree, if not already reported?** (name, department, city, state)	**Institution name** (name, branch or city, state or province, country)	**What is the full name and location of the institution from which you received this degree?** (name, nranch or city, state, country)	✓		✓	✓
	Degree completion	**In what month and year was this degree awarded?**	**Month/year degree granted**	**In what month and year was this degree awarded?**	✓		✓	✓
	Degree start date	*Not collected*	**Month/year degree that you started your degree**	*Not collected*			✓	
	Degree level	**What type of degree did you receive?** -Bachelor's degree (e.g., BS, BA, AB) -Master's degree (e.g., MS, MA, MBA) -Doctorate (e.g., PhD, DSc, EdD) -Other professional degree (e.g., JD, LLB, MD, DDS, DVM) -Other	**Have you received a degree of this type?** -Bachelor's degree (e.g., BA, BS, AB) or equivalent -Master's degree (e.g., MS, MA, MBA) or equivalent -Research doctorate degree -Other	**Thinking about your highest degree, what type of degree is it?** *(If you have more than one doctorate or doctorate-equivalent degree, think about the first such degree you received.)* -Doctorate degree (e.g., PhD, EdD, DSc, Candidate of science, Dr. rer. nat., etc.) -Doctorate-equivalent degree (e.g., MD, DO, DDS, DVM, etc.) -Master's degree (e.g., MS, MA, MPH, Diploma, etc.) -Undergraduate degree (e.g., BS, BA, AB)	✓		✓	✓

Broad Concept	Specific Topic	NSCG/SDR Question Wording	SED Question Wording	ECDS Question Wording	Questionnaire Item Indicator			
					NSCG	SDR	SED	ECDS
First Bachelor's Degree (continued)	Major field of study and second major	**What is the major field of study and second major (if any) for this degree?**	**Primary field of study**	**What was your primary field of study for this degree?**	✓		✓	✓
	Major category code	**Using the FIELD OF STUDY list on pages 20 -21, choose the code that <u>best</u> describes the major field of study and second major (if any) for this degree**	**Field number form list on pages 6-7**	*Not collected*	✓		✓	
Education Financing	Source of financial support	**What type of financial support did you receive to finance any undergraduate or graduate degrees you <u>completed before February 1, 2015</u>?** (Respondents are asked to complete this question separately for their undergraduate degrees educational financing and graduate degrees educational financing) -Financial assistance from parents, spouse, other relatives, <u>not to be repaid</u> -Financial assistance from your employer -Financial assistance from the Veterans Educational Assistance Act (i.e., the G.I. Bill) -Loans from parents or other relatives, <u>to be repaid</u> -Loans from the school you attended, banks, federal or state government -Tuition waivers, fellowships, grants, scholarships -Assistantships or work study -Earnings from employment -Personal savings -Other	**Please indicate whether each of the following was a source of financial support during graduate school.** (Mark Yes or No for each) -Fellowship, scholarship -Grant -Teaching assistantship -Research assistant -Other assistantship -Traineeship -Internship, Clinical residency -Loans (from any source) -Personal savings -Personal earnings during graduate school (other than sources listed above) -Spouse's, partner's, or family's earnings or savings -Employer reimbursement/assistance -Foreign (non-U.S.) support -Other	**Which of the following were your sources of financial support during graduate school?** (Mark all that apply) -Fellowship or scholarship from an institution in the United States -Fellowship or scholarship from abroad -Grant -Teaching assistantship -Research assistantship -Other assistantship -Traineeship -Internship or clinical residency -Loans from any source -Personal savings -Personal earnings during graduate school other than sources already listed -Spouse's, partner's, or family's earnings or savings -Employer reimbursement or assistance -Foreign (non-U.S.) support other than a fellowship or scholarship -Other	✓		✓	✓
		Not collected	**Which TWO sources listed in the previous question provided the most support?**	*Not collected*			✓	
	Tuition remission	*Not collected*	**Did you receive full or partial tuition remission** **(waiver) for your doctoral studies?** (Mark (X) one) -No, I did not receive any tuition remission -Yes, I received less than 1/3 of tuition -Yes, I received between 1/3 and 2/3 of tuition -Yes, I received more than 2/3 of tuition, but less than full -Yes, I received full tuition remission	*Not collected*			✓	

Broad Concept	Specific Topic	NSCG/SDR Question Wording	SED Question Wording	ECDS Question Wording	Questionnaire Item Indicator			
					NSCG	SDR	SED	ECDS
Education Financing (continued)	Amount borrowed and owed	**The next question asks about the TOTAL amount you have borrowed to finance undergraduate and graduate degrees you completed before February 1, 2015, and how much you still owed as of February 1, 2015.** (Respondents are asked to provide borrowed/ owed amounts for total completed undergraduate degrees and borrowed/owed amounts for total completed graduate degrees) -$0 -$1-$10,000 -$10,001-$20,000 -$30,001-$40,000 -$40,001-$50,000 -$50,001-$60,000 -$60,001-$70,000 -$70,001-$80,000 -$80,001-$90,000 -$90,001 or more	**When you receive your doctoral degree, how much money will you owe that is directly related to your undergraduate and graduate education?** (Respondents are asked to provide owed amount for undergraduate education and owed amount for graduate education) -None -$10,000 or less -$10,001-$20,000 -$30,001-$40,000 -$40,001-$50,000 -$50,001-$60,000 -$60,001-$70,000 -$70,001-$80,000 -$80,001-$90,000 -$90,001 or more	**When you receive your doctoral degree, how much money did you owe that was directly related to your undergraduate and graduate education? How much money do you currently owe that is directly related to your undergraduate and graduate education?** (Respondents are asked to provide borrowed/ owed amounts for undergraduate education and borrowed/owed amounts for graduate education) -None -$10,000 or less -$10,001-$20,000 -$30,001-$40,000 -$40,001-$50,000 -$50,001-$60,000 -$60,001-$70,000 -$70,001-$80,000 -$80,001-$90,000 -$90,001 or more	✓		✓	✓
Current Enrollment	Currently enrolled indicator	**During the week of February 1, 2015, were you enrolled in or taking courses at a college or university?**	*Not collected*	**Are you currently enrolled in a doctoral or doctoral-equivalent program?**	✓	✓		✓
	Type of enrollment	*(If Yes)* **Were you taking courses or enrolled as...** (Mark one answer) -a full-time student in a degree program -a part-time student in a degree program -not enrolled in a degree program, but taking courses	*Not collected*	*Not collected*	✓	✓		

Broad Concept	Specific Topic	NSCG/SDR Question Wording	SED Question Wording	ECDS Question Wording	Questionnaire Item Indicator			
					NSCG	SDR	SED	ECDS
Current Enrollment (continued)	Degree level	**Toward what degree were you working?** (Mark one answer) *If you were working toward more than one degree, mark the level for the highest degree.* *-No specific degree* *-Bachelor's degree (e.g., BS, BA, AB)* *-Master's degree (e.g., MS, MA, MBA)* *-Doctorate (e.g., PhD, DSc, EdD)* *-Other professional degree (e.g., JD, LLB, MD, DDS, DVM)* *-Other*	*Not collected*	*Not collected*	✓	✓		
	Major field of study	**What was the primary field of study for this degree?**	*Not collected*	*Not collected*	✓	✓		
	Major category code	**Using the FIELD OF STUDY list on pages 20-21, choose the code that best describes the field of study for this degree**	*Not collected*	*Not collected*	✓			
	Reasons for further education	**For which of the following reasons were you taking classes or enrolled?** (Mark Yes or No for each item) -To gain further education before beginning a career -To prepare for graduate school or further education -To change your academic or occupational field -To gain further skills or knowledge in your academic or occupational field -For licensure or certification -To increase opportunities for promotion, advancement, or higher salary -Required or expected by employer -For leisure or personal interest -Other	*Not collected*	*Not collected*	✓	✓		
	Current enrollment funding	**Were any of your school-related costs for taking courses paid for by an employer?**	*Not collected*	*Not collected*	✓			
Postgraduation Plans	Postgraduation location	*Not collected*	**In what country or state do you intend to live after graduation (within the next year)?**	*Not collected*			✓	

Broad Concept	Specific Topic	NSCG/SDR Question Wording	SED Question Wording	ECDS Question Wording	Questionnaire Item Indicator			
					NSCG	SDR	SED	ECDS
Postgraduation Plans (continued)	Postdoc intentions	*Not collected*	**Do you intend to take a "postdoc" position?** *A "postdoc" is a temporary position primarily for gaining additional education and training in research, usually awarded in academe, industry, government, or a non-profit organization.*	*Not collected*			✓	
	Postgraduate training and/or employment status	*Not collected*	**What is the status of your postgraduate plans (in the next year)?** (Mark one) -Returning to, or continuing in, predoctoral employment -Have signed contract or made definite commitment for a "postdoc" or other work -Negotiating with one or more specific organizations -Seeking position but have no specific prospects -Other full-time degree program (e.g., MD, DDS, JD, MBA, etc.) -Do not plan to work or study (e.g., family commitments, etc.) -Other	*Not collected*			✓	
		Not collected	**What best describes your postgraduate plans (within the next year)?** (Mark one) -"Postdoc" or further training -Employment	*Not collected*			✓	
		Not collected	**What best describes the nature of your further training or study?** (Mark one) -Postdoc" fellowship -Postdoc" research associateship -Traineeship -Internship, clinical residency -Other training	*Not collected*			✓	
		Not collected	**What best describes the nature of your employment?** (Mark one) -Employment (other than military service) -Military service	*Not collected*			✓	

Broad Concept	Specific Topic	NSCG/SDR Question Wording	SED Question Wording	ECDS Question Wording	Questionnaire Item Indicator			
					NSCG	SDR	SED	ECDS
Postgraduation Plans (continued)	Postdoc funding source	*Not collected*	(*If postdoc or further training*) **What will be the main source of financial support for your "postdoc" or further training within the next year?** (Mark one) -U.S. government -Industry/business -College or university -Private foundation -Nonprofit, other than private foundation or college -Foreign government -Other	*Not collected*			✓	
Postgraduation First Principal Job	Employer type	*Not collected*	**What <u>one</u> type of principal employer will you be working for (or training with) in the next year?** ***EDUCATION*** -U.S. 4-year college or university other than medical school -U.S. medical school (including university-affiliated hospital or medical center) -U.S. university-affiliated research institute -U.S. community or 2-year college -U.S. preschool, elementary, middle, secondary school or school system -Foreign educational institution ***GOVERNMENT (other than educational institution)*** -Foreign government -U.S. federal government -U.S. state government -U.S. local government ***PRIVATE SECTOR (other than educational institution)*** -Not-for-profit organization -Industry (for profit) ***OTHER*** -Self-employed	*Not collected*			✓	

Broad Concept	Specific Topic	NSCG/SDR Question Wording	SED Question Wording	ECDS Question Wording	Questionnaire Item Indicator			
					NSCG	SDR	SED	ECDS
Postgraduation First Principal Job (continued)	Employer name and location	*Not collected*	**Please name the organization and geographic location where you will work or study.** (oganization name, state/country) **Is this organization a college or university?**	*Not collected*			✓	
	Salary amount	*Not collected*	**What will be your basic annual salary for this principal job (in the next year)? Do not include bonuses or additional compensation for summertime teaching or research. If you are not salaried, please estimate your earned income**	*Not collected*			✓	
	Salary time period	*Not collected*	**How many months does this salary cover?**	*Not collected*			✓	
	Work activities	*Not collected*	**What will be your primary and secondary work activities?** (Mark one primary activity and one secondary activity) -Research and development -Teaching -Management or administration -Professional services to individuals -Other	*Not collected*			✓	
Demographics-Family Information	Marital status	**On February 1, 2015 were you...** (Mark one answer) -married -living in marriage-like relationship -widowed -separated -divorced -never married	**What is your marital status?** (Mark one answer) -Married -Living in marriage-like relationship -Widowed -Separated -Divorced -Never married	**What is your current marital status?** (Mark one answer) -Married -Living in marriage-like relationship -Never married -Widowed -Separated -Divorced	✓	✓	✓	✓
	Spouse/partner employment status	(*If Married or Living in a marriage-like relationship*) **During the week of February 1, 2015, was your spouse or partner working?** -Yes, fulltime -Yes, parttime -No	*Not collected*	**Does your spouse/partner currently have a paid position?**	✓	✓		✓

Broad Concept	Specific Topic	NSCG/SDR Question Wording	SED Question Wording	ECDS Question Wording	Questionnaire Item Indicator			
					NSCG	SDR	SED	ECDS
Demographics-Family Information (continued)	Spouse/partner employment technical expertise	**Did your spouse's or partner's duties on this job require technical expertise of a bachelor's degree in...** (Mark Yes or No for each item) -engineering, computer science, math, or the natural sciences? -the social sciences? -some other field (e.g., health, business, or education)?	*Not collected*	*Not collected*	✓	✓		
	Dependent children indicator	**As of the week of February 1, 2015 did you have <u>any children</u> living with you as part of your family?**	**Not including yourself or your spouse/partner, how many dependents (children or adults) do you have-that is, how many others receive at least one half of their financial support from you?** (write in the number of dependents in each age category) -5 years of age or younger -6 to 18 years -9 years or older	**Do you have any dependents?**	✓	✓	✓	✓
		(*If Yes*) **How many of these children living with you as part of your family were...** (write in the number of dependents in each age category) -under age 2 -aged 2-5 -aged 6-11 -aged 12-18 -aged 19 or older		(*If Yes*) **How dependents do you have in each of the following age groups.** (write in the number of dependents in each age category) -Aged 5 or younger -Aged 6-18 -Aged 6-11 -Aged 12-18 -Aged 19-25 -Over age 25	✓	✓	✓	✓
	Parent education-mother	**What is the highest level of education completed by your <u>Mother or female guardian</u>?** (Mark one answer) -Less than high school -High school diploma or equivalent -Some college, vocational, or trade school (including 2-year degrees) -Bachelor's degree (e.g., BS, BA, AB) -Master's degree (e.g., MS, MA, MBA) -Professional degree (e.g., JD, LLB, MD, DDS, DVM) -Doctorate (e.g., PhD, DSc, EdD) -Not applicable	**What is the highest educational attainment of your mother?** (Mark one answer) -Less than high school/secondary school graduate -High school/secondary school graduate -Some college -Bachelor's degree -Master's degree (e.g., MA, MS, MBA, MSW, etc.) -Professional degree (e.g., MD, DDS, JD, D.Min, Psy.D, etc.) -Research doctoral degree -Not applicable/Unknown	**What is the highest level of education completed by your m<u>other (stepmother or female guardian)</u>?** (Mark one answer) -Less than high school diploma -High school diploma or equivalent -Some college, vocational, or trade school (including 2-year degrees) -Associate degree (e.g., AA, AS, AAS) -Undergraduate degree (e.g., BS, BA, AB) -Master's degree (e.g., MS, MA, MBA) -Professional degree (e.g., JD, LLB, MD, DDS, DVM) or Doctorate degree (e.g., PhD, DSc, EdD) - Other degree - Don't know	✓		✓	✓

Broad Concept	Specific Topic	NSCG/SDR Question Wording	SED Question Wording	ECDS Question Wording	Questionnaire Item Indicator			
					NSCG	SDR	SED	ECDS
Demographics-Family Information (continued)	Parent education-father	**What is the highest level of education completed by your <u>father or male guardian</u>?** (Mark one answer) -Less than high school -High school diploma or equivalent -Some college, vocational, or trade school (including 2-year degrees) -Bachelor's degree (e.g., BS, BA, AB) -Master's degree (e.g., MS, MA, MBA) -Professional degree (e.g., JD, LLB, MD, DDS, DVM) -Doctorate (e.g., PhD, DSc, EdD) -Not applicable	**What is the highest educational attainment of your father?** (Mark one answer) -Less than high school/secondary school graduate -High school/secondary school graduate -Some college -Bachelor's degree -Master's degree (e.g., MA, MS, MBA, MSW, etc.) -Professional degree (e.g., MD, DDS, JD, D.Min, Psy.D, etc.) -Research doctoral degree -Not applicable/Unknown	**What is the highest level of education completed by your <u>father (stepfather or male guardian)</u>?** (Mark one answer) -Less than high school diploma -High school diploma or equivalent -Some college, vocational, or trade school (including 2-year degrees) -Associate degree (e.g., AA, AS, AAS) -Undergraduate degree (e.g., BS, BA, AB) -Master's degree (e.g., MS, MA, MBA) -Professional degree (e.g., JD, LLB, MD, DDS, DVM) or Doctorate degree (e.g., PhD, DSc, EdD) -Other degree -Don't know	✓		✓	✓
Demographics-Citizenship Information	Residence status	**On February 1, 2015 were you living in the United States, Puerto Rico, or another U.S. territory, or were you living in another country?** (Mark one answer) -United States or Puerto Rico -Another U.S. Territory -Another country	*Not collected*	*Not collected*	✓	✓		
	U.S. citizen indicator	**On February 1, 2015 were you a...** -U.S. citizen -Non-U.S. citizen	**What is your citizenship status?** (Mark one answer) -U.S. CITIZEN -Since birth -Naturalized NON-U.S. CITIZEN -With a Permanent U.S. Resident Visa ("Green Card")? -With a Temporary U.S. Resident Visa?	**Are you a citizen of the United States?** -Yes -No	✓	✓	✓	✓
	U.S. citizen type	(*If U.S. citizen*) **Were you a U.S. citizen...** (Mark one answer) -Born in the United States., Puerto Rico, or another U.S. territory? -Born abroad of U.S. citizen parents? -By naturalization?		(*If U.S. citizen*) **Were you a U.S. citizen...** (Mark one answer) -born in the United States -born in Puerto Rico, Guam, the United States, Virgin Islands, or Northern Marianas -born abroad of American parent parents -by naturalization	✓	✓	✓	✓

Broad Concept	Specific Topic	NSCG/SDR Question Wording	SED Question Wording	ECDS Question Wording	Questionnaire Item Indicator			
					NSCG	SDR	SED	ECDS
Demographics-Citizenship Information (continued)	Non-U.S. citizen type	(*If non-U.S. citizen*) **Were you a non-U.S. citizen...** -with a Permanent U.S. Resident Visa (green card)? -with a Temporary U.S. Resident Visa?		(*If non-U.S. citizen*) **Are you a non-U.S. citizen...** -with a Permanent U.S. Resident Visa (green card)? -with a Temporary U.S. Resident Visa?	✓	✓	✓	✓
	Non-U.S. citizen type-visa type	(*If a Temporary U.S. Resident Visa holder*) **On February 1, 2015 did you hold a visa issued...** (Mark one answer) -For temporary work (e.g., H-1B, L-1A, L-1B, etc.) -For study or training (e.g., F-1, J-1, H-3, etc.) -To you as the dependent of another person (e.g., F-2, H-4, J-2, K-2, L-2, etc.) -For any other reason	*Not collected*	(If a Temporary U.S. Resident Visa holder) **What type of visa do you have?** -H-1B or other H series temporary visa -F-1 or other F series temporary visa -J-1 or other J series temporary visa -TN (Trade NAFTA) temporary visa -Other type of temporary visa	✓			✓
	Country of citizenship	(*If non-U.S. citizen*) **Of which country are you a citizen?**	(*If non-U.S. citizen*) **Of which country are you a citizen?**	**Of which country do you hold citizenship?**	✓	✓	✓	✓
	U.S. year of entry	**In what year did you first come to the United States for 6 months or longer?**	*Not collected*	**In what year did you come to the United States for the first time to stay for a period of 6 months or more?**	✓			✓
	Tenure of U.S. residence	*Not collected*	*Not collected*	**How many years total have you lived in the United States?**				✓
	Previous visa type	*Not collected*	*Not collected*	**Were you already in the United States on a visa before applying for your current position?**				✓
	Initial visit to United States	*Not collected*	*Not collected*	**Before arriving in the United States for the first time, did you plan to return to your home country?**				✓
	Future residence plans	*Not collected*	*Not collected*	**Do you plan to return to your home country when your current position ends?** **Do you plan to return to your home country within the next 10 years?**				✓

Broad Concept	Specific Topic	NSCG/SDR Question Wording	SED Question Wording	ECDS Question Wording	Questionnaire Item Indicator			
					NSCG	SDR	SED	ECDS
Demographics-Citizenship Information (continued)	Factors that influence migration	*Not collected*	*Not collected*	**Did each of the following factors have an influence on your decision to return to your home country?** -Job or economic opportunities in your home country -Educational opportunities in your home country -Family-related reasons -Visa issues -Enjoy living in your home country -Other factors **Did each of the following factors have an influence on your decision to stay in the United States?** -Need for work experience in the United States to advance career in your home country -Career opportunities in the Unted States -Enjoy living in the United States -Economic conditions in your home country -Political situation in your home country -Family-related reasons				✓
	Initial visa type	**What kind of visa did you hold when you first came to the United States for 6 months or longer?** (Mark one answer) -Permanent U.S. Resident Visa (Green Card) -Temporary U.S. Resident Visa for temporary work (e.g., H-1B, L-1A, L-1B, etc.) -Temporary U.S. Resident Visa for study or training (e.g., F-1, J-1, H-3, etc.) -Temporary U.S. Resident Visa as the dependent of another person (e.g., F-2, H-4, J-2, K-2, L-2, etc.)	*Not collected*	*Not collected*	✓			✓

Broad Concept	Specific Topic	NSCG/SDR Question Wording	SED Question Wording	ECDS Question Wording	Questionnaire Item Indicator			
					NSCG	SDR	SED	ECDS
Demographics-Citizenship Information (continued)	Reasons for coming to the United States	**Which factors were important in your decision to first come to the United States for 6 months or longer?** (*Mark Yes or No for each item*) -Family-related reasons -Educational opportunities in the United States -Job or economic opportunities -Scientific or professional infrastructure in my field -Other	*Not collected*	**Which of the following factors influenced your decision to come to the United States?** (*Mark Yes or No for each item*) -Needed a U.S. position to advance career in home country -To work with a particular researcher/individual -To work in a particular department/lab/organization -Career opportunities -Followed the example of other colleagues who moved to the United States -Economic conditions in home country -To get an education in the United States -Family-related reasons -Other	✓			✓
		Which two factors in question E15 were the most important reasons in your decision to come to the United States?	*Not collected*	*Not collected*	✓			
	Dual citizenship indicator	**Are you a citizen of more than one country?**	*Not collected*	*Not collected*	✓			
Basic Demographic Information	Place of birth	**In what U.S. state, U.S. territory, or foreign country were you born?**	*Not collected*	*Not collected*	✓			
	Gender	**Are you…** -male -female	**Are you…** -male -female	**What is your sex?** -Male -Female	✓		✓	✓
	Hispanic indicator	**Are you of Hispanic, Latino, or Spanish origin?** (Mark one answer) -No, not of Hispanic, Latino, or Spanish origin -Yes, Mexican, Mexican American, or Chicano -Yes, Puerto Rican -Yes, Cuban -Yes, another Hispanic, Latino, or Spanish origin	**Are you of Hispanic or Latino?** (Mark one answer) -No, I am not Hispanic or Latino -Yes, I am Mexican or Chicano -Yes, I am Puerto Rican -Yes, I am Cuban -Yes, I am Other Hispanic or Latino	**Are you of Hispanic, Latino, or Spanish origin?** (Mark one answer) -No, not of Hispanic, Latino, or Spanish origin -Yes, Mexican, Mexican American, or Chicano -Yes, Puerto Rican -Yes, Cuban -Yes, South American -Yes, another Hispanic, Latino, or Spanish origin	✓		✓	✓

Broad Concept	Specific Topic	NSCG/SDR Question Wording	SED Question Wording	ECDS Question Wording	Questionnaire Item Indicator			
					NSCG	SDR	SED	ECDS
Basic Demographic Information (Continued)	Race indicator	**What is your race?** (Mark one or more) -American Indian or Alaska Native -Native Hawaiian or other Pacific Islander -Asian -Black or African American -White	**What is your racial background?** (Mark one or more) -American Indian or Alaska Native -Native Hawaiian or other Pacific Islander -Asian -Black or African American -White	**What is your race?** (Mark one or more) -White -Black or African American -Asian -Native Hawaiian or other Pacific Islander -American Indian or Alaska Native	✓		✓	✓
	Birthdate	**What is your birthdate?** (month, day, year)	**What is your date of birth?** (nonth, day, year)	**In what year were you born?**	✓	✓	✓	✓
	Language	*Not collected*	*Not collected*	**Is English your first language?**				✓
Demographics-Disability Status	Sight limitation	**What is the USUAL degree of difficulty you have with SEEING words or letters in ordinary newsprint (with glasses/contact lenses, if you usually wear them)?** (Mark one answer) -None -Slight -Moderate -Severe -Unable to do	**What is the USUAL degree of difficulty you have with SEEING words or letters in ordinary newsprint (with glasses/contact lenses, if you usually wear them)?** (Mark one answer) -None -Slight -Moderate -Severe -Unable to do	**What is the USUAL degree of difficulty you have with SEEING words or letters in ordinary newsprint (with glasses/contact lenses, if you usually wear them)?** (Mark one answer) -None -Slight -Moderate -Severe -Unable to do	✓	✓	✓	✓
	Hearing limitation	**What is the USUAL degree of difficulty you have with HEARING what is normally said in conversation with another person (with hearing aid, if you usually wear one)?** (Mark one answer) -None -Slight -Moderate -Severe -Unable to do	**What is the USUAL degree of difficulty you have with HEARING what is normally said in conversation with another person (with hearing aid, if you usually wear one)?** (Mark one answer) -None -Slight -Moderate -Severe -Unable to do	**What is the USUAL degree of difficulty you have with HEARING what is normally said in conversation with another person (with hearing aid, if you usually wear one)?** (Mark one answer) -None -Slight -Moderate -Severe -Unable to do	✓	✓	✓	✓
	Walking limitation	**What is the USUAL degree of difficulty you have with WALKING without human or mechanical assistance or using stairs?** (Mark one answer) -None -Slight -Moderate -Severe -Unable to do	**What is the USUAL degree of difficulty you have with WALKING without human or mechanical assistance or using stairs?** (Mark one answer) -None -Slight -Moderate -Severe -Unable to do	**What is the USUAL degree of difficulty you have with WALKING without human or mechanical assistance or using stairs?** (Mark one answer) -None -Slight -Moderate -Severe -Unable to do	✓	✓	✓	✓

Broad Concept	Specific Topic	NSCG/SDR Question Wording	SED Question Wording	ECDS Question Wording	Questionnaire Item Indicator			
					NSCG	SDR	SED	ECDS
Demographics-Disability Status (continued)	Lifting limitation	**What is the USUAL degree of difficulty you have with LIFTING or carrying something as heavy as 10 pounds, such as a bag of groceries?** (Mark one answer) -None -Slight -Moderate -Severe -Unable to do	**What is the USUAL degree of difficulty you have with LIFTING or carrying something as heavy as 10 pounds, such as a bag of groceries?** (Mark one answer) -None -Slight -Moderate -Severe -Unable to do	**What is the USUAL degree of difficulty you have with LIFTING or carrying something as heavy as 10 pounds, such as a bag of groceries?** (Mark one answer) -None -Slight -Moderate -Severe -Unable to do	✓	✓	✓	✓
	Cognitive limitation	**What is the USUAL degree of difficulty you have with CONCENTRATING, REMEMBERING, or MAKING DECISIONS because of a physical, mental, or emotional condition?** (Mark one answer) -None -Slight -Moderate -Severe	**What is the USUAL degree of difficulty you have with CONCENTRATING, REMEMBERING, or MAKING DECISIONS because of a physical, mental, or emotional condition?** (Mark one answer) -None -Slight -Moderate -Severe	**What is the USUAL degree of difficulty you have with CONCENTRATING, REMEMBERING, or MAKING DECISIONS because of a physical, mental, or emotional condition?** (Mark one answer) -None -Slight -Moderate -Severe	✓	✓	✓	✓
	Disability-Age of onset	**What is the earliest age at which you first began experiencing any difficulties in any of these areas?**	**What is the earliest age at which you first began experiencing any difficulties in any of these areas?**	**What is the earliest age at which you first began experiencing any difficulties in any of these areas?**	✓	✓	✓	✓
Survey Data Collection Information	Social Security Number	*Not collected*	**Please fill in the last four digits of your Social**	*Not collected*			✓	
	Mode preference	**How would you like to complete future rounds of this survey?** (Mark one answer) -A questionnaire sent in the mail -A web questionnaire on the Internet -A telephone interview -No preference	*Not collected*	**How do you prefer to participate in research surveys?** -On a personal computer or laptop -On a tablet -Telephone -Mail -In-Person	✓	✓		✓

NOTES: ECDS = Early Career Doctorates Survey; NSCG = National Survey of College Graduates; SDR = Survey of Doctorate Recipients; SED = Survey of Earned Doctorates.
SOURCE: Table provided by NCSES.

Appendix C

Workshop on Data Needs Agenda

The National Academies of Sciences, Engineering, and Medicine
National Academy of Sciences Building, Members Room
2101 Constitution Ave., NW

OCTOBER 20, THURSDAY

DATA USER WORKSHOP—PUBLIC SESSION

8:30-9:00 **WELCOME AND INTRODUCTIONS**
Rita Colwell, *Cochair, Johns Hopkins University and University of Maryland*
James House, *Cochair, University of Michigan*
John Gawalt, *National Center for Science and Engineering Statistics, National Science Foundation*

9:00-10:35 **SESSION I: POLICY, INDUSTRY, AND GOVERNMENT DATA NEEDS**
MODERATOR: Jennifer Sue Bond, *Council on Competitiveness*
PANELISTS:
William Bonvillian, *Massachusetts Institute of Technology*
Lisa Frehill, *Energetics Technology Center (currently IPA at NSF)*
William Harris, *Science Foundation Arizona*
Brian Yoder, *American Society for Engineering Education*

Tobin Smith, *Association of American Universities*
Leigh Ann Pennington, *Oak Ridge Institute for Science Education*

10:35-10:50 COFFEE BREAK

10:50-12:00 SESSION II: ACADEMIC RESEARCH DATA NEEDS
MODERATOR: Willie Pearson, *Georgia Institute of Technology*
PANELISTS:
Kaye Husbands Fealing, *Georgia Institute of Technology*
Michael Teitelbaum, *Harvard University*
Donna Ginther, *University of Kansas*
John Skrentny, *University of California, San Diego*

12:00-1:00 WORKING LUNCH TO CONTINUE MORNING DISCUSSION

1:00-2:30 SESSION III: INTERNATIONAL PERSPECTIVES
MODERATOR: Maresi Nerad, *University of Washington*
PANELISTS:
Fernando Galindo-Rueda, *OECD, France*
Beate Scholz, *Consultant (via WebEx from Germany)*
Janet Metcalfe, *Vitae, United Kingdom*
Jack Gambino, *Statistics Canada*

2:30-2:50 COFFEE BREAK

2:50-4:10 SESSION IV: EARLY CAREER PERSPECTIVES
MODERATOR: Geoff Davis, *Verily*
PANELISTS:
Michael Roach, *Cornell University*
Yu Tao, *Stevens Institute of Technology*
Andrea Stith, *BioFrontiers, University of Colorado Boulder*
Neil Ruiz, *George Washington University*
Prasanna Tambe, *New York University*

4:10-5:00 FLOOR DISCUSSION AND WRAP-UP
Rita Colwell, *Cochair, Johns Hopkins University and University of Maryland*

Appendix D

Potential Uses of Unemployment Insurance Earnings Data to Enhance Longitudinal Data

THE LONGITUDINAL PROBLEM

One of the important data needs identified by the panel is to enable researchers to trace the employment of the nation's science and engineering workforce over time. The NCSES human resources data program has several "intake" surveys that capture essential data on these individuals, including their institutions, programs of study, demographics, motivations, and attitudes, among other domains. At intake, a great deal is known about these people, and NCSES regularly collects data on new cohorts entering the science and engineering workforce. Thus the data system is a powerful resource for tracking flows into the science and engineering workforce (with the exception of immigrants educated abroad who enter the United States with similar qualifications). The Survey of Earned Doctorates (SED) and the National Survey of College Graduates (NSCG) collect data from a significant number of these highly educated people. The SED then feeds into the Survey of Doctorate Recipients (SDR), which has a longitudinal component. Because of the size of the sample, however, the SDR reduces the size of the panel component to maintain coverage breadth for its first wave.

The excellent data on flows into the science and engineering workforce constitute a good starting point. However, the careers of scientists and engineers change and end, and individuals may emigrate either permanently or temporarily. Decades ago, employment arrangements for workers in science and engineering fields were very stable—as was the case for a great many white-collar workers. The average number of career jobs held by persons in the latter years of the baby boom (1957–1964) was 11.7 from age 18 to 48 (Angeles, 2016). Nearly one-half of these jobs were held from ages 18 to 24. Among newer entrants to the science and engineering workforce in recent years, similar job churning is likely, especially shortly after graduation, when workers are trying to place themselves in employment arrangements they find satisfactory, or at least as good as they can get.

UNEMPLOYMENT INSURANCE EARNINGS DATA AS A SUPPLEMENT TO THE CORE SCIENTISTS AND ENGINEERS STATISTICAL DATA SYSTEM (SESTAT) EFFORT

All states collect earnings data for covered employment[1] for their unemployment insurance programs. This system collects quarterly earnings, the employer identification number (EIN), and the number of weeks employed for covered workers. With the EIN, one can recover the North American Industry Classification System number, which gives the type of industry in which the worker is employed. These data can be used to compute the person's wage and salary income, mobility between employers, and location of the employer.[2] With the EIN one can also recover the employer's wage bill and the number of persons it employs. There are no occupational data. The unemployment insurance earnings data are highly comparable across states (the Census Bureau has done the work of compiling all these state data into a single structure) and include data on *all* jobs held, regardless of state or locality, that are covered by the unemployment insurance system. The data do not include self-employment or other income, but in the grand scheme of work for pay, the coverage is excellent, and the sources of income that are missing are missing consistently through time.

The Census Bureau has agreements with all states allowing it to use the unemployment insurance earnings data to match to census surveys. Matching the unemployment insurance earnings data to other databases would require either a new batch of memoranda of understanding with the various states or new legislation. However, these challenges are reduced for the NSCG because the data are collected by the Census Bureau, and the legal authority to match is in place. This would mean, however, that the newly matched data would be covered by Title 13. The large size of the NSCG makes the use of administrative data matching compelling if only on cost grounds.

With this matching, one could track employment, earnings, and location for the science and engineering workforce survey respondents

---

[1]Uncovered employment, except for the self-employed, will be infrequent for SESTAT respondents.

[2]Although not necessarily the location of the worker.

surveyed by the Census Bureau. This would provide reliable employment and earnings data for every respondent surveyed by the Census Bureau, and at a per-case cost orders of magnitude less than that of the current survey effort. If, at intake, respondents provided their cell phone number, then whenever a respondent changed employers, one could ask a few questions that would illuminate the job transition, such as what the survey respondent's new occupation is and why the person left his or her previous job. Just a handful of questions would provide timely information about employment transitions. Respondents might be contacted with a few questions once every 5 years or so, with administrative data matching enabling a more passive tracking of their careers. This inexpensive tracking strategy would produce a longitudinal file that would enable analysis of career dynamics for a broader range of respondents than is currently possible. The unemployment insurance earnings data are quarterly, so the employment and earnings history information obtainable from administrative data is more detailed in terms of both quality and frequency relative to what can be attempted in the SDR.[3]

TRADEOFFS

This administrative effort would not replace the current intake questionnaires, but supplement them. Even absent any survey effort directed at past entrants in the NCSES data system of scientist and engineers, matching in the unemployment insurance earnings data would provide significant information on the evolution of science and engineering careers as well as degree holders in other disciplines, and would do so at very modest cost. A "skinny" survey effort that asked a few trenchant questions, such as the individual's occupation and whether his or her responsibilities are more managerial, entrepreneurial, or research and development in nature would get to the heart of what many want to know about the nation's science and engineering human capital infrastructure. Making such abbreviated contact every 5 years or when the subject changed employers would reveal a great deal about what the nation's stock of science and engineering workers are doing relative to research and development.

What would be lost would be the far lengthier reinterviews for prior survey intake cases. However, these reinterviews do not dominate unemployment insurance earnings data when it comes to understanding job transitions, employment, earnings, and industry. This modification would change altogether the way employment dynamics in the survey datasets are viewed

[3]A data source that provides monthly rather than quarterly data is the National Directory of New Hires, a database that provides information on people taking new jobs. See https://www.acf.hhs.gov/css/resource/overview-of-national-directory-of-new-hires [January 2018].

and support examination of these dynamics across a wider variety of demographics and fields of study and a greater number of persons tracked relative to what could be afforded with the current strategy.

Appendix E

Biographical Sketches of Panel Members and Staff

RITA R. COLWELL (*Cochair*) is distinguished university professor in the Department of Cell Biology and Molecular Genetics at the University of Maryland and at the Johns Hopkins University Bloomberg School of Public Health. She served as director of the National Science Foundation (NSF) between 1998 and 2004. Before joining NSF, she was president of the University of Maryland Biotechnology Institute and professor of microbiology and biotechnology at the University of Maryland. She was also a member of the National Science Board from 1984 to 1990. Her interests include K–12 science and mathematics education, graduate science and engineering education, and the increased participation of women and minorities in science and engineering. She is a member of the National Academy of Sciences, the Royal Swedish Academy of Sciences, the Royal Society of Canada, the American Academy of Arts and Sciences, and the American Philosophical Society. She has been awarded 61 honorary degrees from institutions of higher education, and is the recipient of the 2017 National Science Board Vannevar Bush Award, the 2010 Stockholm Water Prize, and the 2006 National Medal of Science. She holds a Ph.D. in oceanography from the University of Washington.

JAMES S. HOUSE (*Cochair*) is Angus Campbell distinguished university professor emeritus of survey research, public policy, and sociology at the University of Michigan. He is a former director of the Survey Research Center at the University of Michigan Institute for Social Research. Prior to joining the University of Michigan, he taught at Duke University and the University of North Carolina at Chapel Hill. His research has focused on topics in the areas of social psychology, political sociology, social structure and personality, psychosocial and socioeconomic factors in health, survey research methods, and American society. He is an elected member of the American Academy of Arts and Sciences, the National Academy of Sciences, and the Institute of Medicine (now National Academy of Medicine). He also serves as a member on the Committee on National Statistics. He holds a Ph.D. in social psychology from the University of Michigan.

JENNIFER SUE BOND is senior fellow at the Council on Competitiveness and a consultant for science and technology policy issues. She is the former director of the Science and Engineering Indicators Program at the National Science Foundation. In the past, she also served as a senior international policy analyst at the White House Office of Science and Technology Policy; as a congressional fellow in the Senate, focused on education and science policy issues; and an advisor to the Battelle Center for Science and Technology Policy. For more than a decade she was a U.S. representative to OECD and vice chair of OECD's National Experts on Science, Technology, and Industry Indicators. She worked with the European Union on several issues, including the Community Innovation Surveys, Public Attitudes Towards Science and Technology, and Mega-Science Issues. She holds an M.A. in science, technology, and public policy from the George Washington University.

GEOFFREY DAVIS is staff quantitative analyst on the Google Accelerated Science team at Alphabet in Mountain View, California. His previous Alphabet roles include Verily (Google Life Sciences), Google Infrastructure, Google AdWords, and Google News. Prior to joining Google, he was assistant professor in the Mathematics Department at Dartmouth College, Texas Instruments visiting assistant professor in the Electrical and Computer Engineering Department at Rice University, a researcher in the Signal Processing Group at Microsoft Research, a developer at the San Francisco-based startup company 4charity, a visiting scholar at Sigma Xi: The Scientific Research Society, and a Werthheim fellow in the Labor and Worklife Program at Harvard Law School. He has had a long-standing interest in science education and policy issues and is a past member of the Science and Engineering Workforce Project of the National Bureau of Economic Research. He is a recipient of the Leon K. Kirchmayer Prize Paper Award of the Institute of Electrical and Electronics Engineers. He holds a Ph.D. in applied mathematics from the Courant Institute at New York University.

DONALD A. DILLMAN is Regents professor in the Department of Sociology at Washington State University. He also serves as deputy director for research and development in the Social and Economic Sciences Research Center at Washington State University. Previously, he served as senior survey methodologist in the office of the director at the U.S. Census Bureau. He is recognized internationally as a major contributor to the development of modern mail, telephone, and Internet survey methods. In 2000, he received the Roger Herriot Award for Innovation in Federal Statistics for his work at the Census Bureau. He is a fellow of the American Association for the Advancement of Science and the American Statistical Association, and served as past president of the American Association for Public Opinion Research and the Rural Sociological Society. He also serves as a member of the Committee on National Statistics. He holds a Ph.D. in sociology from Iowa State University.

RICHARD B. FREEMAN is Herbert Ascherman chair in economics at Harvard University. He is currently serving as faculty co-chair of the Harvard University Trade Union Program. He is also director of the Labor Studies Program at the National Bureau of Economic Research (NBER), senior research fellow in labor markets at the London School of Economics' Centre for Economic Performance, and visiting professor at the London School of Economics (LSE). He also directs the NBER/Sloan Science Engineering Workforce Project and an LSE research program on the effects of the Internet on labor markets, social behavior, and the economy. His many research interests include the job market for scientists and engineers. He is the recipient of the Mincer Lifetime Achievement Prize from the Society of Labor Economics and the IZA Prize in Labor Economics. He was also appointed Frances Perkins fellow of the American Academy of Political and Social Science. He holds a Ph.D. in economics from Harvard University.

JACK G. GAMBINO is director of the Household Survey Methods Division at Statistics Canada. He manages a division consisting of three subdivisions responsible for the design, development, implementation, and maintenance of almost all of Statistics Canada's household surveys. Previously, he led a division that was responsible for the methodology of Canada's census of population. He participated in two redesigns of the Canadian labor force survey, proposing innovations to improve the survey design and planning studies to evaluate these proposals. He also provided guidance and advice to methodologists and client program managers on methodological issues, including research and development of innovative methods, and chaired the Methodology Research and Development Committee, which was responsible for overseeing research in the methodology branch. He received his Ph.D. in statistics from the University of Toronto.

KRISZTINA MARTON (*Study Director*) is a senior program officer with the Committee on National Statistics (CNSTAT) of the National Academies of Sciences, Engineering, and Medicine. She served as study director for numerous CNSTAT consensus panels, standing committees, workshops, and expert meetings, including the Panel on Addressing Priority Technical Issues for the American Community Survey, the Panel on Statistical Methods for Measuring the Group Quarters Population in the American Community Survey, and the Panel on Redesigning the Commercial Buildings and Residential Energy Consumption Surveys of the Energy Information Administration. Prior to joining CNSTAT, she was a survey researcher at Mathematica Policy Research and a survey director in the Ohio State University Center for Survey Research. She holds a Ph.D. in communication with an interdisciplinary specialization in survey research from Ohio State University.

MARESI NERAD is professor of higher education leadership in the College of Education and founding director of the Center for Innovation and Research in Graduate Education (CIRGE) at the University of Washington. Previously, she was director of graduate research at the University of California, Berkeley, where she undertook various studies on graduate education in the United States and abroad, including examinations of factors affecting time-to-degree and doctoral attrition, as well as a national study on the career paths of doctorates. In her current work at CIRGE, she oversees research on the quality of doctoral programs, careers of doctorate holders, and evaluations of innovations in doctoral education in the United States and in an international context. She served on many national and institutional doctoral education reviews and on U.S. and international advisory boards, such as the International Advisory Committee for Science and Engineering of the National Science Foundation; the Presidential Innovation Board of the University of Bremen, Germany; and the Graduate Academy of the Goethe University of Frankfurt. She holds a Ph.D. in higher education from the University of California, Berkeley.

RANDALL J. OLSEN is professor emeritus of economics and former director of the Center for Human Resource Research at Ohio State University. He also served as director of the Initiative in Population Research, an interdisciplinary population center at Ohio State University. For more than 20 years he has been heavily involved with the National Longitudinal Surveys (NLS), a set of large-scale surveys that follow cohorts of Americans to collect data on labor force behavior, educational attainment, training, and related topics. He has overseen the transition of the NLS from legacy data collection systems to an integrated system for handling all phases of survey work, from instrument authoring through data dissemination. His interests are in econometrics, labor economics, and applied micro and economic demography. He has served as an associate editor for *Evaluation Review*, the *Journal of the American Statistical Association*, and *Demography*. He holds a Ph.D. in economics from the University of Chicago.

WILLIE PEARSON, Jr. is professor of sociology and former chair of the School of History, Technology, and Society at Georgia Institute of Technology. Previously, he held a distinguished appointment as Wake Forest professor of sociology at Wake Forest University and an adjunct position in medical education at Wake Forest University School of Medicine. Most of his research has centered on the career patterns of Ph.D. scientists, particularly African Americans, and on human resource issues in science and engineering. He received the Giants in Science award from the Quality Education for Minorities in Mathematics, Science, and Engineering Network; is a national associate of the National Academies of Sciences, Engineering, and Medicine; and is a fellow of the American Association for the Advancement of Science. He holds a Ph.D. in sociology from Southern Illinois University at Carbondale.

KEITH F. RUST is vice president and associate director of the statistical staff at Westat, Inc. He is also a research associate professor in the Joint Program on Survey Methodology at the University of Maryland. Previously, he worked for the Australian Bureau of Statistics. He has extensive experience in sampling methods, the design and specification of large-scale sample surveys, and analysis of survey data. His areas of special expertise include methods for analyzing large, complex datasets; methods of deriving survey weights; and sampling error estimation procedures. He has applied his research and knowledge to a variety of education research projects over the past several years, including the Programme for International Student Assessment and the National Assessment of Educational Progress. He is a fellow of the American Statistical Association and a member of the International Statistical Institute and served as a member of the Committee on National Statistics. He received a Ph.D. in biostatistics from the University of Michigan.

NORA CATE SCHAEFFER is Sewell Bascom professor of sociology in the Department of Sociology at the University of Wisconsin–Madison and faculty director of the University of Wisconsin Survey Center, where she teaches courses in survey research methods and conducts research on questionnaire design and interaction during survey interviews. She serves on the editorial boards of *Sociological Methods and Research* and *Public Opinion Quarterly,* and is also a member of the General Social Survey Board of Overseers. She recently completed terms as Council on Sections representative for the Survey Research Methods Section of the American Statistical Association and as a member of the Census Advisory Committee of Professional Associations. She was selected as a fellow of the American Statistical Association. She previously served on the Committee on National Statistics. She holds a Ph.D. in sociology from the University of Chicago.

JAMES WAGNER is research associate professor at the Survey Research Center and research affiliate at the Population Studies Center, both at the Institute for Social Research, University of Michigan. He teaches in the Joint Program in Survey Methodology and serves as associate editor of *Survey Research Methods*. He also serves as principal investigator on several studies and is the chief mathematical statistician on the National Survey of Family Growth. His research interests include nonresponse error, quality indicators for survey data, and responsive or adaptive design. He also develops and implements responsive design plans for large surveys. He holds a Ph.D. from the program in survey methodology at the University in Michigan.

YU XIE is Bert G. Kerstetter '66 university professor of sociology and director of the Center on Contemporary China at the Princeton Institute of International and Regional Studies, Princeton University. He is also visiting chair professor of the Center for Social Research, Peking University, in China. Previously, he was distinguished university professor of sociology, statistics, and public policy at the University of Michigan. His main areas of interest are social stratification, demography, statistical methods, Chinese studies, and the sociology of science. He authored the book *Is American Science in Decline?* with Alexandra Killewald. He is a member of the American Academy of Arts and Sciences, Academia Sinica, and the National Academy of Sciences. He holds a Ph.D. in sociology from the University of Wisconsin–Madison.

COMMITTEE ON NATIONAL STATISTICS

The Committee on National Statistics was established in 1972 at the National Academies of Sciences, Engineering, and Medicine to improve the statistical methods and information on which public policy decisions are based. The committee carries out studies, workshops, and other activities to foster better measures and fuller understanding of the economy, the environment, public health, crime, education, immigration, poverty, welfare, and other public policy issues. It also evaluates ongoing statistical programs and tracks the statistical policy and coordinating activities of the federal government, serving a unique role at the intersection of statistics and public policy. The committee's work is supported by a consortium of federal agencies through a National Science Foundation grant, a National Agricultural Statistics Service cooperative agreement, and several individual contracts.